Stem Cell Bioprocessing and Manufacturing

Stem Cell Bioprocessing and Manufacturing

Editors

Joaquim M. S. Cabral
Cláudia Lobato da Silva
Maria Margarida Diogo

MDPI • Basel • Beijing • Wuhan • Barcelona • Belgrade • Manchester • Tokyo • Cluj • Tianjin

Editors
Joaquim M. S. Cabral
Universidade de Lisboa
Portugal

Cláudia Lobato da Silva
Universidade de Lisboa
Portugal

Maria Margarida Diogo
Universidade de Lisboa
Portugal

Editorial Office
MDPI
St. Alban-Anlage 66
4052 Basel, Switzerland

This is a reprint of articles from the Special Issue published online in the open access journal *Bioengineering* (ISSN 2306-5354) (available at: https://www.mdpi.com/journal/bioengineering/special_issues/stem_cell_bioprocess).

For citation purposes, cite each article independently as indicated on the article page online and as indicated below:

LastName, A.A.; LastName, B.B.; LastName, C.C. Article Title. *Journal Name* **Year**, *Article Number*, Page Range.

ISBN 978-3-03943-038-3 (Hbk)
ISBN 978-3-03943-039-0 (PDF)

Contents

About the Editors . . . . **vii**

Joaquim M.S. Cabral, Cláudia Lobato da Silva and Maria Margarida Diogo
Stem Cell Bioprocessing and Manufacturing
Reprinted from: *Bioengineering* **2020**, *7*, 84, doi:10.3390/bioengineering7030084 . . . . **1**

Kathleen Van Beylen, Ali Youssef, Alberto Peña Fernández, Toon Lambrechts, Ioannis Papantoniou and Jean-Marie Aerts
Lactate-Based Model Predictive Control Strategy of Cell Growth for Cell Therapy Applications
Reprinted from: *Bioengineering* **2020**, *7*, 78, doi:10.3390/bioengineering7030078 . . . . **5**

Valentin Jossen, Francesco Muoio, Stefano Panella, Yves Harder, Tiziano Tallone and Regine Eibl
An Approach towards a GMP Compliant In-Vitro Expansion of Human Adipose Stem Cells for Autologous Therapies
Reprinted from: *Bioengineering* **2020**, *7*, 77, doi:10.3390/bioengineering7030077 . . . . **23**

Sandra M. Jonsdottir-Buch, Kristbjorg Gunnarsdottir and Olafur E. Sigurjonsson
Human Embryonic-Derived Mesenchymal Progenitor Cells (hES-MP Cells) are Fully Supported in Culture with Human Platelet Lysates
Reprinted from: *Bioengineering* **2020**, *7*, 75, doi:10.3390/bioengineering7030075 . . . . **47**

Josephine Lembong, Robert Kirian, Joseph D. Takacs, Timothy R. Olsen, Lye Theng Lock, Jon A. Rowley and Tabassum Ahsan
Bioreactor Parameters for Microcarrier-Based Human MSC Expansion under Xeno-Free Conditions in a Vertical-Wheel System
Reprinted from: *Bioengineering* **2020**, *7*, 73, doi:10.3390/bioengineering7030073 . . . . **61**

Katharina M. Prautsch, Lucas Degrugillier, Dirk J. Schaefer, Raphael Guzman, Daniel F. Kalbermatten and Srinivas Madduri
Ex-Vivo Stimulation of Adipose Stem Cells by Growth Factors and Fibrin-Hydrogel Assisted Delivery Strategies for Treating Nerve Gap-Injuries
Reprinted from: *Bioengineering* **2020**, *7*, 42, doi:10.3390/bioengineering7020042 . . . . **77**

Rachael Wood, Pelin Durali and Ivan Wall
Impact of Dual Cell Co-culture and Cell-conditioned Media on Yield and Function of a Human Olfactory Cell Line for Regenerative Medicine
Reprinted from: *Bioengineering* **2020**, *7*, 37, doi:10.3390/bioengineering7020037 . . . . **93**

Brian Lee, Breanna S. Borys, Michael S. Kallos, Carlos A. V. Rodrigues, Teresa P. Silva and Joaquim M. S. Cabral
Challenges and Solutions for Commercial Scale Manufacturing of Allogeneic Pluripotent Stem Cell Products
Reprinted from: *Bioengineering* **2020**, *7*, 31, doi:10.3390/bioengineering7020031 . . . . **109**

Nasim Nosoudi, Anson Oommen Jacob, Savannah Stultz, Micah Jordan, Seba Aldabel, Chandra Hohne, James Mosser, Bailey Archacki, Alliah Turner and Paul Turner
Electrospinning Live Cells Using Gelatin and Pullulan
Reprinted from: *Bioengineering* **2020**, *7*, 21, doi:10.3390/bioengineering7010021 . . . . **119**

Sam L. Francis, Angela Yao and Peter F. M. Choong
Culture Time Needed to Scale up Infrapatellar Fat Pad Derived Stem Cells for Cartilage Regeneration: A Systematic Review
Reprinted from: *Bioengineering* **2020**, *7*, 69, doi:10.3390/bioengineering7030069 . . . . **131**

João P. Cotovio and Tiago G. Fernandes
Production of Human Pluripotent Stem Cell-Derived Hepatic Cell Lineages and Liver Organoids: Current Status and Potential Applications
Reprinted from: *Bioengineering* **2020**, *7*, 36, doi:10.3390/bioengineering7020036 . . . . **141**

About the Editors

Joaquim M. S. Cabral Department of Bioengineering and iBB - Institute of Bioengineering and Biosciences, Instituto Superior Técnico, Universidade de Lisboa, Av. Rovisco Pais, Lisboa 1049-001, Portugal. Interests: stem cells research for tissue engineering and regenerative medicine; stem cell bioprocessing and manufacturing: development of novel stem cell bioreactors and advanced bioseparation and purification processes.

Cláudia Lobato da Silva Department of Bioengineering and iBB - Institute of Bioengineering and Biosciences, Instituto Superior Técnico, Universidade de Lisboa, Av. Rovisco Pais, Lisboa 1049-001, Portugal. Interests: the ex-vivo expansion of human stem cells; cellular therapies with human adult stem cells; isolation and purification of stem cells; bioreactors for stem cell culture.

Maria Margarida Diogo Department of Bioengineering and iBB - Institute of Bioengineering and Biosciences, Instituto Superior Técnico, Universidade de Lisboa, Av. Rovisco Pais, Lisboa 1049-001, Portugal. Interests: the development of scalable platforms for highly controlled expansion and differentiation of human pluripotent stem cells (hPSC), embryonic (hESC), induced pluripotent stem cells (hiPSC); stem cell-based purification strategies; regenerative medicine; drug screening; disease modelling.

Editorial

Stem Cell Bioprocessing and Manufacturing

Joaquim M.S. Cabral *, Cláudia Lobato da Silva and Maria Margarida Diogo

Department of Bioengineering and iBB—Institute of Bioengineering and Biosciences, Instituto Superior Técnico, Universidade de Lisboa, Av. Rovisco Pais, 1049-001 Lisboa, Portugal; claudia_lobato@ist.utl.pt (C.L.d.S.); margarida.diogo@ist.utl.pt (M.M.D.)

* Correspondence: joaquim.cabral@ist.utl.pt

Received: 27 July 2020; Accepted: 28 July 2020; Published: 31 July 2020

The next healthcare revolution will apply regenerative medicines using human cells and tissues. Regenerative medicine aims to create biological therapies or in vitro substitutes for the replacement or restoration of tissue function in vivo lost due to failure or disease. However, whilst science has revealed the biomedical potential of this approach, and early products have demonstrated the power of such therapies, there is a need for the development of bioprocess technology for the successful transfer of the laboratory-based practice of stem cell and tissue culture to the clinic as therapeutics through the application of engineering principles and practices. This Special Issue of *Bioengineering* on "Stem Cell Bioprocessing and Manufacturing" addresses the central role in defining the engineering sciences of cell-based therapies by bringing together contributions from worldwide experts on stem cell science and engineering, bioreactor design and bioprocess development, scale-up, and the manufacturing of stem cell-based therapies.

In the last few years, human pluripotent stem cell (hPSC) derivatives have emerged as promising allogeneic cell therapy products, with amazing potential to treat a wide variety of diseases and a vast number of patients globally. Brian Lee and co-authors [1] addressed various challenges related to the manufacturing of PSCs in large quantities for commercialization, which include bioreactor process development—namely, scalable bioreactor technology for the large-scale manufacturing of high-quality therapeutic PSCs derivatives.

Among the most promising hPSC derivatives, hepatic cell lineages represent a potential cell source, holding great potential for biomedical applications, such as in liver cell therapy, disease modelling, and drug discovery. João Cotovio and Tiago Fernandes [2] assessed the production of different hepatic cell lineages from PSCs, including hepatocytes, as well as the emerging strategies to generate hPSC-derived liver organoids, highlighting their current biomedical applications.

Alongside the development of novel bioreactor configurations for cell therapy manufacturing, efforts have also been undertaken to optimize bioreactor operating conditions. In their study focused on the manufacturing of mesenchymal stem/stromal cell (MSC) therapies, Josephine Lembong and colleagues [3] developed a fed-batch, microcarrier-based process in a Vertical-Wheel system, which enhanced media productivity while driving a cost-effective and less labor-intensive cell expansion process. As another strategy to improve the cost-effectiveness of cell manufacturing processes, Kathleen Van Beylen and colleagues [4] developed a lactate-based model predictive control strategy for cell growth monitoring and the control of cell proliferation, by adapting the feeding strategy based on lactate measurements, while envisaging the reduction in unnecessary costs, a particularly relevant issue in large-scale cell manufacturing.

An additional key aspect towards the successful translation of cell therapy products is the need to use animal origin-free products (i.e., xeno(geneic)-free) for the derivation, expansion, and differentiation of stem cells in order to minimize the risks of animal-transmitted diseases and immune reactions to foreign proteins. In this context, Valentin Jossen and colleagues [5] developed a bioprocess approach for the expansion of adipose-derived stem cells (ASC), targeting autologous therapies by employing xeno- and serum-free culture conditions and testing static, planar (2D), and dynamically mixed (3D) cultivation systems. To this end, the authors compared the donor variability in both culture systems and developed a mathematical growth model to describe cell growth, nutrient consumption, and metabolite production. Following this same trend, Sandra M. Jonsdottir-Buch and colleagues [6] described the successful proliferation of mesenchymal progenitors derived from human embryonic stem cells (hES-MP) using a culture medium supplemented with human platelet lysates. These hES-MP cells are proposed as interesting alternatives to adult MSC, and the authors demonstrated that these cells can be grown using platelet lysates, maintaining similar proliferation and differentiation profiles to those expanded in culture medium supplemented with FBS.

In addition to the increasing demand for large-scale cell manufacturing protocols, there is a critical need to establish potency assays for stem cell therapy products and their derivatives. Katharina M. Prautsch and colleagues [7] developed a strategy to improve the potency of ASC for nerve regeneration through ex vivo stimulation of ASC with nerve growth factor (NGF). The authors found that the secretome from NGF-stimulated ASC promoted significant axonal outgrowth in an in vitro setting. Upon in vivo delivery of these stimulated ASC (on fibrin-hydrogel nerve conduits), there was an enhancement of early nerve regeneration in a sciatic nerve gap-injury. For other cell therapy candidates, efforts have continued towards the development of the most appropriate culture conditions to establish regenerative phenotypes. This is the case for olfactory ensheathing cells (OECs), a promising therapy candidate for neuronal tissue repair. Rachael Wood and colleagues [8] showed that neither dual co-culture nor fibroblast-conditioned media support the regenerative human OEC phenotype, which means that the appropriate priming conditions to drive a regenerative phenotype in human OECs are yet to be determined.

Experimental culture conditions are critical for the ex vivo expansion and differentiation of stem cells. In fact, variables such as culture supplements, the purity of the initial cell population, the initial cell concentration, and the duration of culture affect the outcome of stem cell cultures and, consequently, the regenerative potential of ex vivo cultured stem cell-derived products. Sam L. Francis and co-authors [9] focused on the manufacturing of human ASC for articular cartilage regeneration. The authors revealed that there is a higher amount of fat tissue, stromal vascular fraction cell count, and overall yield associated with open (arthrotomy) compared to arthroscopic IFP harvest and described a novel framework for the culture time needed to scale-up the manufacturing of these cells based on the harvesting method.

Finally, cell-delivery methods are a key part of regenerative medicine. The delivery of stem cells and their derivatives can be performed as scaffold-free products (e.g., single cell suspension) or combined with polymer scaffolds. Traditionally, cells and biological agents are implanted into the matrix of the scaffold following electrospinning. The study performed by Nasim Nosoudi and colleagues [10] focused on the development of a novel design that simultaneously introduces cells into the scaffold during the electrospinning process. By demonstrating that human ASC can be directly incorporated into the electrospinning process, maintaining a high viability, the authors suggest the potential benefits of this strategy within the tissue engineering field.

Conflicts of Interest: The authors declare no conflict of interest.

References

1. Lee, B.; Borys, B.S.; Kallos, M.S.; Rodrigues, C.A.V.; Silva, T.P.; Cabral, J.M.S. Challenges and Solutions for Commercial Scale Manufacturing of Allogeneic Pluripotent Stem Cell Products. *Bioengineering* **2020**, *7*, 31. [CrossRef] [PubMed]
2. Cotovio, J.P.; Fernandes, T.G. Production of Human Pluripotent Stem Cell-Derived Hepatic Cell Lineages and Liver Organoids: Current Status and Potential Applications. *Bioengineering* **2020**, *7*, 36. [CrossRef] [PubMed]
3. Lembong, J.; Kirian, R.; Takacs, J.D.; Olsen, T.R.; Lock, L.T.; Rowley, J.A.; Ahsan, T. Bioreactor Parameters for Microcarrier-Based Human MSC Expansion under Xeno-Free Conditions in a Vertical-Wheel System. *Bioengineering* **2020**, *7*, 73. [CrossRef]

4. Van Beylen, K.; Youssef, A.; Peña Fernández, A.; Lambrechts, T.; Papantoniou, I.; Aerts, J.-M. Lactate-Based Model Predictive Control Strategy of Cell Growth for Cell Therapy Applications. *Bioengineering* **2020**, *7*, 78. [CrossRef] [PubMed]
5. Jossen, V.; Muoio, F.; Panella, S.; Harder, Y.; Tallone, T.; Eibl, R. An Approach towards a GMP Compliant In-Vitro Expansion of Human Adipose Stem Cells for Autologous Therapies. *Bioengineering* **2020**, *7*, 77. [CrossRef]
6. Jonsdottir-Buch, S.M.; Gunnarsdottir, K.; Sigurjonsson, O.E. Human Embryonic-Derived Mesenchymal Progenitor Cells (hES-MP Cells) are Fully Supported in Culture with Human Platelet Lysates. *Bioengineering* **2020**, *7*, 75. [CrossRef] [PubMed]
7. Prautsch, K.M.; Degrugillier, L.; Schaefer, D.J.; Guzman, R.; Kalbermatten, D.F.; Madduri, S. Ex-Vivo Stimulation of Adipose Stem Cells by Growth Factors and Fibrin-Hydrogel Assisted Delivery Strategies for Treating Nerve Gap-Injuries. *Bioengineering* **2020**, *7*, 42. [CrossRef] [PubMed]
8. Wood, R.; Durali, P.; Wall, I. Impact of Dual Cell Co-culture and Cell-conditioned Media on Yield and Function of a Human Olfactory Cell Line for Regenerative Medicine. *Bioengineering* **2020**, *7*, 37. [CrossRef] [PubMed]
9. Francis, S.L.; Yao, A.; Choong, P.F.M. Culture Time Needed to Scale up Infrapatellar Fat Pad Derived Stem Cells for Cartilage Regeneration: A Systematic Review. *Bioengineering* **2020**, *7*, 69. [CrossRef] [PubMed]
10. Nosoudi, N.; Oommen, A.J.; Stultz, S.; Jordan, M.; Aldabel, S.; Hohne, C.; Mosser, J.; Archacki, B.; Turner, A.; Turner, P. Electrospinning Live Cells Using Gelatin and Pullulan. *Bioengineering* **2020**, *7*, 21. [CrossRef] [PubMed]

Article

Lactate-Based Model Predictive Control Strategy of Cell Growth for Cell Therapy Applications

Kathleen Van Beylen [1,2], Ali Youssef [1], Alberto Peña Fernández [1], Toon Lambrechts [1,2], Ioannis Papantoniou [2,3,4] and Jean-Marie Aerts [1,*]

1 Department of Biosystems, Division Animal and Human Health Engineering, M3-BIORES: Measure, Model & Manage Bioresponses Laboratory, KU Leuven, Kasteelpark Arenberg 30, 3001 Heverlee, Belgium; kathleen.vanbeylen@kuleuven.be (K.V.B.); ali.youssef@kuleuven.be (A.Y.); alberto.penafernandez@kuleuven.be (A.P.F.); toon.lambrechts@mycellhub.com (T.L.)

2 Prometheus, Division of Skeletal Tissue Engineering, KU Leuven, Onderwijs en Navorsing 1, Herestraat 49, 3000 Leuven, Belgium; ioannis.papantoniou@kuleuven.be

3 Skeletal Biology and Engineering Research Centre, Onderwijs en Navorsing 1, Herestraat 49, 3000 Leuven, Belgium

4 Institute of Chemical Engineering Sciences, Foundation for Research and Technology—Hellas (FORTH), 26504 Patras, Greece

* Correspondence: jean-marie.aerts@kuleuven.be

Received: 30 May 2020; Accepted: 15 July 2020; Published: 20 July 2020

Abstract: Implementing a personalised feeding strategy for each individual batch of a bioprocess could significantly reduce the unnecessary costs of overfeeding the cells. This paper uses lactate measurements during the cell culture process as an indication of cell growth to adapt the feeding strategy accordingly. For this purpose, a model predictive control is used to follow this a priori determined reference trajectory of cumulative lactate. Human progenitor cells from three different donors, which were cultivated in 12-well plates for five days using six different feeding strategies, are used as references. Each experimental set-up is performed in triplicate and for each run an individualised model-based predictive control (MPC) controller is developed. All process models exhibit an accuracy of 99.80% ± 0.02%, and all simulations to reproduce each experimental run, using the data as a reference trajectory, reached their target with a 98.64% ± 0.10% accuracy on average. This work represents a promising framework to control the cell growth through adapting the feeding strategy based on lactate measurements.

Keywords: model predictive control; bio-process; cell growth; lactate; advanced therapy medicinal products

---

1. Introduction

Cell-based products receiving market approval are increasing over the last years. The European Medicine Agency (EMA) has approved 14 medicinal products based on gene therapies, cell therapies or tissue engineering, also called advanced therapies for the European market [1]. The U.S. Food and Drug Administration (FDA) has approved 17 cellular or gene therapy products [2]. Compared to other pharmaceuticals such as small molecule drugs or biologics, the active pharmaceutical ingredient (API) of these cell-based therapies is living cells. An example of such a cell-based therapy is chimeric antigen receptor (CAR) T-cell therapy, where the patient is injected with human immune cells, which are modified to target cancer cells [3]. Another type of cell-based therapy is skeletal tissue engineering, where a cell-based implant is used to regenerate cartilage or bone in the patient instead of using a prosthetic implant, which has the disadvantage that it will need to be replaced within 10–15 years [4]. Besides being the active component of the final medicinal product, cells can also be used as a tool in the manufacturing process to produce the final product, such as extracellular vesicles [5].

With the introduction of this promising group of cell-based or cell-derived products, the necessity to transform the emerging cell therapy and regenerative medicine industrial sector towards a BioPharma 4.0 sector is growing. This revolution should build on a strong IT infrastructure combined with automation technologies in order to use continuous data to steer and optimise bioprocesses in real-time without the need for human interaction [6]. Closely monitoring and controlling the bioprocess tackles the challenge of irreproducible manufacturing processes that are often seen for (personalised) cell-based therapies. This bioprocess variability is inherent to donor variability, the time-varying characteristics of progenitor cells (such as phenotype) and the complexity of living systems [7].

Progenitor cell expansion is a crucial process step whereby clinically relevant numbers are produced typically ranging between 5×10^7–10^8 [8,9]. Currently, progenitor cell expansion relies on fixed protocols which do not take into account the particularities of the cell type, donor characteristics or the batch, leading to suboptimal outcomes [10]. In order to reduce this variability, the process has to remain within predefined boundaries during the whole production process, which is possible by actively adapting critical process parameters (CPP) during the cell expansion process, based on the characteristics and individual needs of a batch. The retuning of the process parameters should be done in a way that would enable the process to follow a predefined (reference) trajectory, providing optimal conditions for the cultured cells. Due to the inherent variability of cells and the time-varying dynamics of the process, modelling and controlling the cell growth is challenging [11].

Active control of cell culture bioprocesses will also result in lower batch-to-batch variability. Without any monitoring or control of cell culture, there could be a high amount of batch rejections due to results of in-process or finished product testing falling out of the predefined boundaries of the validated process. These specifications are described in quality control documents approved by health authorities and are set to assure product quality and safety. The amount of "out of specifications" test results of two different commercial cell therapies was recently described in the biologics license application (BLA) submission of Kymriah® and Yescarta®. Novartis reported 7% and 9% manufacturing failures for Kymriah batches, whereas Kite reported 1% for Yescarta batches [12]. Novartis disclosed that all out of specifications (OOS) results were caused by viability problems, resulting in final products with a viability lower than 80%. The challenge lies in the nature of cell products having an inherent variability and complexity.

Therefore, in this work, a model-based predictive control (MPC) system is proposed as a potential solution to the aforementioned challenges of inherent variability and time-varying dynamics of the cell culture process [13]. MPC exhibits several interesting features, such as intuitive concepts, easy tuning and the ability to control a range of simple and complex phenomena, including systems with time delays, non-minimum phase dynamics, dead times, multivariable cases and instability [14]. While dealing with all these challenges, the MPC can easily incorporate constraints and tailor formulated control objectives [12,13]. Model predictive control offers several important advantages: (1) the process model captures the dynamic and static interactions between input, output and disturbance variables; (2) constraints on inputs and outputs are considered in a systematic manner in the cost function and (3) accurate model predictions can provide early warnings of potential problems [13].

Several studies have investigated the benefits of controlling the environment of cell culture vessels such as dissolved oxygen tension (dO_2), temperature and CO_2 [15]. Instead of using these standard physicochemical process parameters to control the bioprocess, this paper will develop a method to control the metabolic responses of the cells. This metabolic response is measured off-line and is used as an indication of the cell growth, which can only be measured at the end of the bioprocess of adherent cells. An interesting metabolic response to use as an indirect measure for cell growth in a high glucose medium is the cumulative lactate production of the cells over the culture period. Using lactate measures has the advantage, in an environment with excess amount of glucose, that the ratio between lactate production and glucose consumption is a known value (two) based on the anaerobic glycolysis pathway [16]. In high glucose environments, measurements of glucose have a low sensitivity compared to lactate. Lactate concentrations are low in fresh medium and are produced by the cells, resulting in higher sensitivity and indication of whether or not cells are alive. Another advantage is controlling the pH, since this is related to the lactate concentration [17,18]. The control of this pH is important because an increase in extracellular acidosis, i.e., a value below 6.7, leads to a higher amount of apoptosis [19,20].

Furthermore, lowering the lactate concentration by replacing the media for 100%, 50% or 0% of the total working volume has been reported to have a significant effect on the cell growth [15].

The aim of this paper is to describe a framework for controlling process parameters of the cell expansion process based on lactate measurements in combination with a model predictive control approach. As a proof of concept we used lactate measures, but depending on the considered application, the input and output could be chosen differently, taking into account specific process parameters and quality attributes. For example, in low glucose environments, it would be interesting to change the measurement to glucose. By controlling the process parameters, the cell growth can be directed towards a predefined reference trajectory. This research demonstrated the intended goal using experimental data in combination with control strategy simulations.

2. Materials and Methods

2.1. Cell Culture Experiments

In order to develop this framework, we performed experiments on human periosteum-derived cells (hPDCs) and studied their metabolic responses during their cell expansion process. Cell proliferation was the aimed output. This cell growth was represented here by the cumulative lactate produced by the cells. As an input to control the cell growth, we investigated the effect of the total amount of replaced medium.

2.1.1. Cell Culture

The hPDCs used in this study were obtained from periosteal biopsies with patients' informed consent. The performed biopsy procedures, as described by [21], were approved by the Ethics Committee for Human Medical Research (KU Leuven). These cells were expanded until passage 4 and frozen. Culture medium consisted of high glucose Dulbecco's modified Eagle's medium (DMEM + GlutaMAX™ + pyruvate, Gibco™ by Thermo Fisher Scientific, Waltham, MA, USA), supplemented with 10% (v/v) heparin-free pooled human platelet lysate (Stemulate™ by Cook Regentec, Indianapolis, IN, USA) and 1% antibiotic-antimycotic (Gibco™ by Thermo Fisher Scientific).

The cell culture experiment started by thawing three frozen vials, each containing 1 million hPDC cells from a different donor. The cells from these three donors were seeded in three different T175 flask at passage 5 with 27 mL culture medium and incubated in a humidified atmosphere of 90% at 37 °C and 5% CO_2. The culture medium used during the experiment was DMEM supplemented with only 7.5% hPL instead of 10%, which was used for general cell culture expansion and storage. The reason for lowering the amount of hPL is based on knowledge from previous experiments, indicating cells cultured in 7.5% hPL as the condition with the lowest medium cost per population doubling (data not included). Cells were subjected to a 100% medium replacement on day 2 and harvested on day 4 with TrypLE (Gibco™ by Thermo Fisher Scientific). This passaging was repeated once again, with the same seeding density of 5700 cells·cm^{-2}.

2.1.2. Experimental Set-Up

Cells were harvested after the second expansion step and seeded into 6 different 12-well plates (72 wells), each well with a density of 3300 cells·cm^{-2} in 1 mL of DMEM medium supplemented with 7.5% hPL. Reducing the seeding density from the previous 5700 cells·cm^{-2}, which was used for expanding and storing of cells, to 3300 cells·cm^{-2} was, on the one hand, based on previous experiments. These experiments indicated a seeding density of 3300 cells·cm^{-2} to be a more cost-effective use of the culture vessel, due to a lower population doubling time and similar cell number harvested at the end of the cell culture. On the other hand, a lower seeding density would also provide more cell culture time before reaching 80% of confluency, resulting in a higher amount of input and output data points. The cells were cultured during 5 days while the medium was replaced according to 6 different medium replacement strategies, as indicated in Table 1.

Table 1. Overview of medium replacement strategies. The amount of medium replaced is indicated as a percentage of the total working volume of the well, which changed over the different days.

Medium Replaced		**Day**				
		1	**2**	**3**	**4**	**Explanation**
Condition	**1**	10.0	10.0	25.0	45.0	Increasing: 0; 15; 20
	2	12.5	25.0	36.5	50.0	A steady increase of 12.5
	3	10.0	20.0	35.0	45.0	A decreasing increase: 10; 15; 10
	4	10.0	10.0	10.0	10.0	Constant replacement
	5	50.0	50.0	50.0	50.0	Constant replacement
	6	100.0	100.0	100.0	100.0	Constant replacement

All conditions were performed for three different donors in triplicates (54 wells). In addition a control condition was set up in each of the six 12-well plates in triplicates (18 wells), which had the same medium replacement scheme as condition 6, but the cells were from a pool of the three different donors to account for possible well plate differences.

2.1.3. Lactate Measurements and Cell Counts

During the 5 days of cell culture, 100 µL medium samples were taken every day from each and stored at −80 °C. Therefore, a minimum of 10% medium replacement was required. The medium samples were analysed for lactate with the CEDEX medium analyser (Roche, Custom Biotech, Belgium) after thawing. After five days of cell culture expansion, the cells were harvested using TrypLE express and counted with trypan blue 0.25% using a Bürker haemocytometer.

2.2. Model-Based Control and Optimisation

2.2.1. System Identification and Modelling

The main goal of this work is to (1) optimise the cell proliferation, combined with (2) minimising the use of medium, which can be achieved by tuning a process parameter to steer the process towards a defined growth trajectory. In order to solve this optimisation problem, a model-based predictive control (MPC) approach is used, which is shown in Figure 1.

The control strategy consists of a dynamic model to forecast the future behaviour of the system (predicted outputs $\hat{y}(k+N_p|k)$, at time k with prediction horizon N_p). This predictive knowledge is used in combination with the past knowledge of previous input and output measurements of the system and a reference trajectory ($r(k+N_p)$) to calculate the future errors ($\hat{e}(k+N_p|k)$). The optimiser will take these errors in to account, together with the cost function (J) and the constraints, to formulate the optimal control decision (future inputs $\hat{u}(k+N_c|k)$, estimated at time k with control horizon N_c) to be used as inputs to minimise the deviation from the reference trajectory [23].

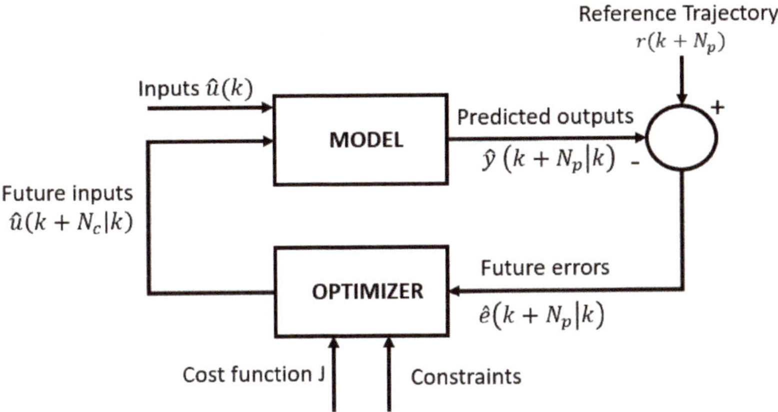

Figure 1. Model predictive controller scheme [22].

A first step in developing a model-based controller is to develop a model of the process. When no readily available mechanistic model or knowledge is available, a model can be identified based on measuring process inputs and outputs. Several methods can be used, but an approach that has been proven successful in many applications is system identification. This approach assumes that the observed input–output relations of the system are the manifestation of the dominant processes occurring within the system under study. Typically, a transfer function (TF) model structure is estimated as an objective and the parsimonious mathematical description of the process is considered [24].

The reason for using a data-based model predictive controller is based on the multiple advantages it has regarding controlling and optimising systems compared to classical proportional–integral–derivative (PID) controllers [25]. The model will predict the lactate increase and use time varying parameters combined with an a priori defined reference trajectory required for the complex and time-varying nature of the cells. Furthermore, the model is able to include feedback knowledge of experiments and extract the main processes to see the effect on the growth. In addition, it can take into account constraints on the input and output variables, use short prediction horizons and avoid time delay problems.

2.2.2. Interpolated Data

One of the challenges faced during the present study was the sparsity of the data points, with only one data point every 24 h. Therefore, an interpolation step was needed, for which the method of piecewise linear interpolation was used. In order to do this, all collected data points are used and the data in between are estimated using a linear function [26]. For a dataset of n points (t_1, y_1), .., (t_n, y_n) with $t_1 < t_n$, the piecewise linear interpolation for point t situated at $t_k < t < t_{k+1}$, is described by

$$\mathrm{y(t)} = y_k + \frac{y_{k+1} - y_k}{t_{k+1} - t_k} \cdot (t - t_k), \quad (1)$$

where y (mmol) is again the accumulated lactate produced and t (days) is the culture period in days. The values (t_k, y_k) and (t_{k+1}, y_{k+1}) are collected data points, whereas $(t, \mathrm{y(t)})$ is an interpolated data point. The resulted interpolated data were used as a reference trajectory in the simulation step for the developed model predictive controller.

2.2.3. Prediction Model

The MPC approach requires a dynamic model which forecasts the output, in this case the cell growth. Furthermore, the model relates the process parameters, used as inputs, to this desired output.

The goal of this work is to estimate the growth of the cells during the cell culturing phase. However, since adherent cells cannot be measured directly in this phase, an indirect measure of cell growth is used, namely, accumulated lactate produced by the cells during proliferation.

The advantage of the previous mentioned system-identification methods, such as transfer function models, is that they develop the process models directly based on measured process data and thus can take into account differences between cell types and/or time-varying characteristics.

Figure 2 shows a representation of the lactate concentrations over time, with medium replacements at certain time points k.

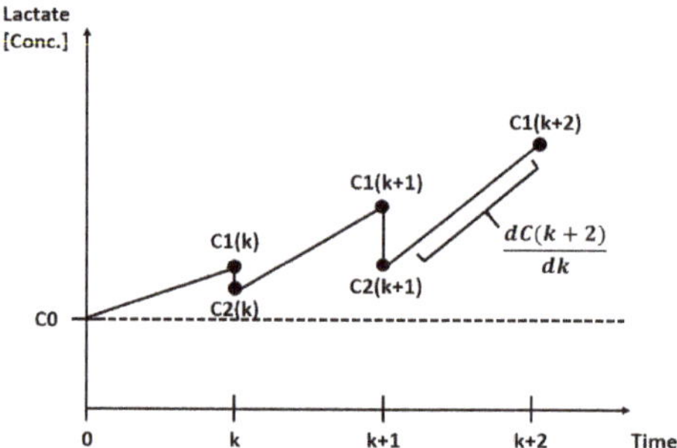

Figure 2. Simplified lactate concentration graph during a cell expansion period with medium replacements at time points k, $k+1$ *and* $k+2$.

At time zero, the cell culture has an initial lactate concentration which is equal to the concentration in fresh medium and is called the baseline concentration C_0. While the cells proliferate, they consume nutrients such as glucose and produce waste products such as lactate. Therefore, the lactate concentration increases between time zero and k from the initial C_0 until $C_1(k)$. At time k, the medium is replaced with $U(k)$ as a percentage of the working volume of the vessel. After medium replacement, the lactate concentration $C_1(k)$ decreases to $C_2(k)$ as described in the following equation:

$$C_2(k) = C_1(k) - C_1(k)U(k) + C_0U(k). \tag{2}$$

To control this lactate production, the amount of medium used to replenish the cells can be used as the manipulated process parameter (or control input).

A data-based mechanistic model approach was used to describe the effect of changing the medium replacement on the cumulative lactate production. A transfer function input–output model structure is used for system model identification, as little knowledge of the complex cell behaviour is required a priori.

In this research, dynamic auto-regressive exogenous (DARX) variables are estimated using the CAPTAIN toolbox [27] in MATLAB version 2018b. The DARX model is used in the analysis to allow a changing relation between medium replacement and accumulated lactate during the cell culture period [28]. The model structure is described as follows [22,23]:

$$y_t = \frac{B(z^{-1},t)}{A(z^{-1},t)}u_{t-\delta} + \frac{1}{A(z^{-1},t)}e_t, \tag{3}$$

where y_t is the output (accumulated lactate (mmol)) of the system and $u_{t-\delta}$ the input (accumulated medium replaced (mL) with a certain time delay δ. The additive noise e_t is assumed to have a zero mean and uncorrelated variance $N\left(0,\sigma^2\right)$. The series A and B have time varying parameters described by the following equations:

$$A\left(z^{-1},t\right) = 1 + a_{1,t}z^{-1} + a_{2,t}z^{-2} + \cdots + a_{n,t}z^{-n_a} \quad (4)$$

$$B\left(z^{-1},t\right) = b_{0,t} + b_{1,t}z^{-1} + b_{2,t}z^{-2} + \cdots + b_{m,t}z^{-n_b}, \quad (5)$$

where the backward shift operator z^{-1}, applied on the model parameters $a_{i,t}$ and $b_{i,t}$, can also be expressed as:

$$a_{i,t}z^{-i} = a_{i,t}(t-i). \quad (6)$$

To obtain the relation between input and output, estimated by the polynomials A and B, experiments were performed. These experiments changed the process parameter (u, medium replacement) while measuring the effect on the output (y, cumulative lactate concentration). The model parameters were estimated using refined instrumental variable (RIV) algorithms [27]. The most suitable reduced order model structure was selected based on two identification criteria, namely, the coefficient of determination R^2 and Young identification criterion (YIC). The orders of these polynomials in Equations (4) and (5) are n_a and n_b. For these data, model orders between 1 and 2 for n and m respectively were evaluated, including time delays between 0 and 1. The best fit was obtained using first order polynomials with a fixed a_1 parameter in time and a variable b_0 during all the different time points. The accuracy of this fit is measured with MATLAB version 2018b using normalised root mean square error (NRMSE) with the goodness of fit function. This method is described as follows:

$$\text{NRMSE} = 1 - \frac{\left\|y_{ref} - y_{fit}\right\|}{\left\|y_{ref} - mean\left(y_{ref}\right)\right\|} \quad (7)$$

where y_{fit}, the modelled data, compared to y_{ref}, the reference data. The NRMSE equals 1 for a perfect fit.

2.2.4. Cost Function

The optimal process parameter values are those which steer the system towards the reference trajectory function. These values are calculated as the ones minimising a controller's cost function. This cost function consists of one term to minimise the difference between the predicted output ($\hat{y}$) and the reference trajectory (r), and another term to minimise the change of the control signal (Δu) (i.e., the replaced medium volume). This equation is as follows:

$$J(Nc, Np) = \sum_{j=1}^{Np} \delta(j)\left[\hat{y}(k+j|k) - r(k+j)\right]^2 + \sum_{j=1}^{Nc} \lambda(j)[\Delta u(k+j-1)]^2, \quad (8)$$

where Nc is the control horizon, Np is the prediction horizon (time points where y is controlled to follow r) and δ, λ are used as weights to create a relevance ranking [22,29].

2.2.5. Constraints

The solutions to optimise the system are subject to constraints. The input, manipulated to control the system, could be restricted by physical boundaries. For example, replacing the medium for 100% in certain vessels is impossible without the risk of removing cells together with the medium. In addition, the output of the system could also be restricted to assure product quality, feasibility or safety. For example, the lactate concentration of the cell culture system is limited to avoid toxic lactate levels, meaning a value of 20 mM [30]. There was no need to implement these constraints in the current work, since the toxic lactate threshold was never reached in these experiments, not even for the condition of minimal lactate replacement.

2.2.6. Simulation

In this paper, the use of a model-based predictive control approach was evaluated for cell growth control by quantifying the performance of the controller based on simulated control actions. More specifically, the experimental data were used to identify time-varying transfer function models describing the dynamic relations between cumulative lactate concentrations and medium refreshments and these models were used in combination with the control algorithms to simulate the needed medium refreshments. The reference trajectory for cumulative lactate concentration was assumed to be the cumulative lactate concentrations actually measured for each condition.

3. Results

3.1. Collected Data

Figure 3 shows the amount of accumulated lactate produced and the cell number after the five days of cell expansion. These results are summarised in Table 2, and show the average of the triplicates for the different donors and different medium replacement conditions, which were explained in Table 1.

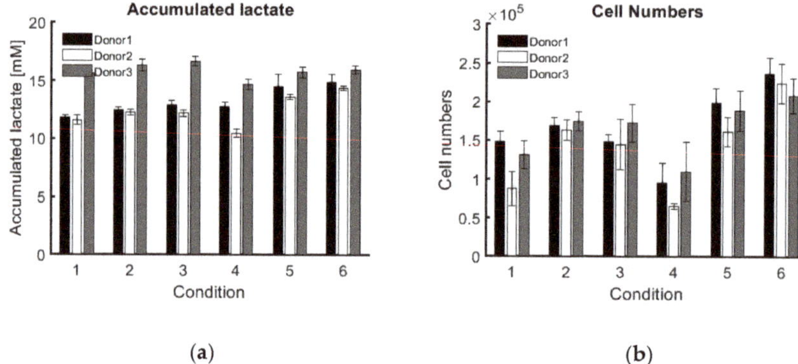

(**a**) (**b**)

Figure 3. Average results with error bars for the triplicate experiments of each individual donor with each specific medium replacement strategy. (**a**) Total accumulated lactate produced after 5 days of cell culture (mM). (**b**) Cell numbers counted at the end of the cell culture period.

Table 2. Total amount of cells harvested after the cell culture expansion, which is averaged over the triplicates for each donor and condition (100,000 cells).

Harvested Cells		**Donor**		
		1	**2**	**3**
Condition	1	1.48 ± 0.13	0.88 ± 0.22	1.31 ± 0.18
	2	1.69 ± 0.11	1.63 ± 0.13	1.75 ± 0.12
	3	1.48 ± 0.10	1.45 ± 0.33	1.73 ± 0.24
	4	0.96 ± 0.25	0.65 ± 0.04	1.10 ± 0.38
	5	1.99 ± 0.19	1.61 ± 0.18	1.89 ± 0.26
	6	2.37 ± 0.21	2.24 ± 0.25	2.08 ± 0.22

Figure 4 represents an example of measured lactate concentration over time, similar to Figure 2, but adapted to the data from the experiments with donor 1, condition 5 and triplicate 1.

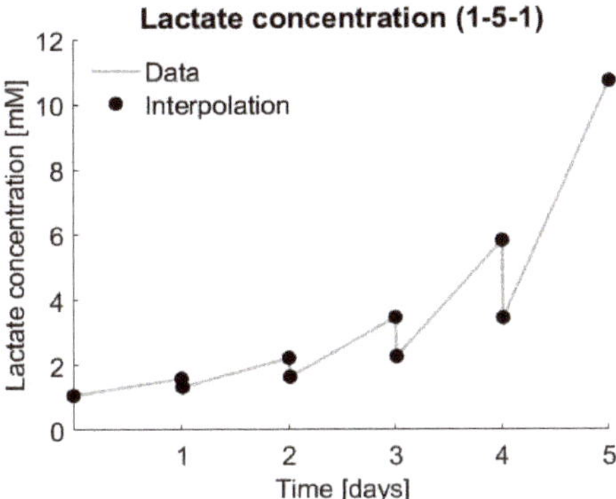

Figure 4. Lactate concentration over time (days) for the results of donor 1, condition 5 and triplicate 1.

Table 3 represents how efficiently the amount of medium is used for proliferation by the cells. This is calculated by dividing the total amount of cells by the total amount of medium supplied during the cell expansion. The table indicates that the most efficient medium replacement strategy, meaning the most cells per amount of medium used, is donor- and not method-dependent. All three different donors require different medium replacement strategies. However, giving the cells the highest amount of medium (condition 6 with 100% medium replacement every 24 h) always results in the highest amount of cells at the end of the expansion. Also, the lowest amount of medium replacement always results in the lowest amount of cells at the end of the expansion.

Table 3. Efficiency of medium used over the total cell culture period, calculated by dividing the total cell numbers by the total amount of medium used (10,000 cells mL^{-1}).

Medium Efficiency		**Donor**		
		1	**2**	**3**
Condition	**1**	7.79 ± 0.71	4.61 ± 1.15	6.91 ± 0.93
	2	7.53 ± 0.48	7.29 ± 0.60	7.79 ± 0.55
	3	7.06 ± 0.45	6.88 ± 1.55	8.21 ± 1.15
	4	6.85 ± 1.79	4.67 ± 0.29	7.86 ± 2.70
	5	6.64 ± 0.63	5.38 ± 0.61	6.29 ± 0.88
	6	4.73 ± 0.41	4.48 ± 0.51	4.17 ± 0.44

Therefore, to develop a model predictive controller, it is always necessary to keep in mind what the goal or reference is. If the goal is to predict the feeding strategy of the cells in order to reach the highest amount of cells, the controller would suggest to replace the medium as much as possible. The downside is that resources are wasted due to unnecessary medium replacements. A more interesting question would be to ask the controller how much medium should be replaced to reach, for example, 80% of the total amount of cells according to a maximum medium replacement strategy (condition 6) in the same amount of time. Or another question could be, in a case where a patient has a procedure scheduled in fixed amount of time, e.g., eight weeks: how much medium should be provided to the cells to reach the therapeutically-required amount of cells in eight weeks?

Table 4 represents the results of the average amount of lactate produced by each cell at the end of the cell expansion. This relation is interesting for translating the accumulated amount of lactate produced to the amount of cells. However, this number differs for each donor and differs even more between different medium-replacement strategies. Condition 6, in which the medium is replaced 100% every day, could be a representation of how the cells produce lactate in an optimal environment. Condition 4, in which only 10% of the medium is replaced every day, has a significantly higher amount of lactate produced over the expansion period, which is due to either a lack of nutrients and growth factors, or inhibiting factors such as lactate itself.

Table 4. The accumulated lactate divided by the cell numbers at the end of the cell culture period and divided by total culture time (120 h), averaged over the triplicates for each donor and condition (10,000,000 mM cell^{-1} h^{-1}).

Lactate Production		Donor		
		1	2	3
Condition	1	6.71 ± 0.65	11.49 ± 2.84	10.07 ± 1.55
	2	6.16 ± 0.33	6.29 ± 0.65	7.83 ± 0.67
	3	7.27 ± 0.66	7.26 ± 1.65	8.18 ± 1.38
	4	11.78 ± 3.93	13.35 ± 0.88	12.09 ± 4.21
	5	6.14 ± 0.99	7.09 ± 0.89	7.06 ± 1.11
	6	5.27 ± 0.40	5.40 ± 0.69	6.44 ± 0.65

One of the biological reasons for this difference in lactate produced by cells could be that cells die due to this lactate inhibition or nutrient and growth factor depletion. Therefore, less cells are counted in the end than actually lived, causing a higher lactate·cell^{-1} ratio [31]. Another reason could be that cells are changing their metabolic profiles [32].

3.2. Interpolated Data

When using the piecewise linear interpolation method, the fit was 100%, since all data points are being used. The piecewise linear interpolation methods were further used to interpolate the sparse data set. Instead of using only one data point every day, the data are interpolated to one data point every hour, which reflects a more realistic approach for field conditions.

3.3. Prediction Model

The model parameters for the DARX model, represented in Equation (3), were obtained using first order polynomials with a fixed a_1 parameter (cf. Equation (4)) in time and a variable b_0 (cf. Equation (5)) during all the different time points. The accuracy of this DARX model compared to the piecewise interpolated output is shown in Table 5 and visualised in Figure 5.

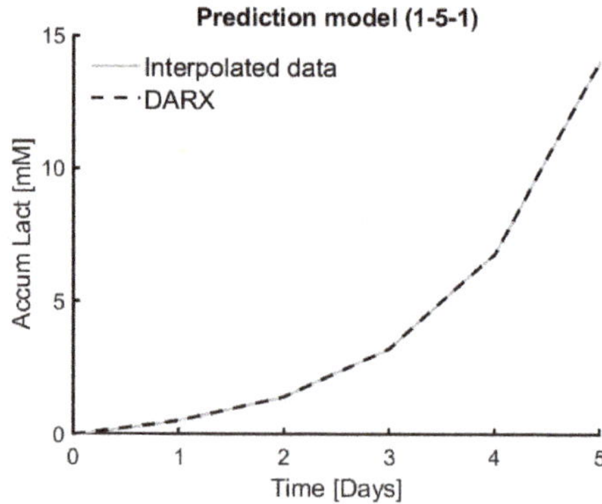

Figure 5. Visualisation of dynamic auto-regressive exogenous (DARX) model for accumalated lactate (mM) compared to the interpolated data for the experiment of donor 1, condition 5 and triplicate 1.

Table 5. Accuracy measured using normalised root mean square error (NRMSE) of DARX model compared to output of the interpolated data for each condition, donor and triplicates. The DARX model used is either the one based on the corresponding experimental data (diagonal values) or on the data of an experimental triplicate. The value is NRMSE multiplied by 100 and expressed as a percentage, with 100 being a perfect fit.

NRMSE		Donor 1			Donor 2			Donor 3		
		DARX Model Triplicate								
Condition	Experimental data triplicate	1	2	3	1	2	3	1	2	3
1	1	99.76	96.17	94.72	99.77	97.11	95.63	99.79	93.83	98.15
	2	96.28	99.77	91.74	97.17	99.78	93.34	94.08	99.79	92.77
	3	94.65	91.42	99.76	95.50	93.00	99.78	98.12	92.45	99.79
2	1	99.78	93.27	97.14	99.80	95.59	98.14	99.82	91.97	94.21
	2	93.56	99.79	95.87	95.73	99.80	94.17	92.42	99.80	97.16
	3	97.21	95.78	99.79	98.06	93.94	99.78	94.47	97.13	99.80
3	1	99.79	93.34	98.64	99.80	98.30	95.47	99.82	98.02	93.74
	2	93.66	99.79	92.78	98.29	99.80	95.19	97.99	99.82	94.68
	3	98.67	92.39	99.80	95.34	95.02	99.79	93.49	94.55	99.81
4	1	99.75	99.13	92.60	99.75	97.45	95.98	99.77	90.54	95.41
	2	99.22	99.77	93.26	97.46	99.74	93.62	91.10	99.77	95.41
	3	92.20	92.93	99.77	95.87	93.31	99.76	95.55	95.26	99.77
5	1	99.81	98.30	87.19	99.82	98.54	95.71	99.83	94.76	91.61
	2	98.31	99.81	86.28	98.54	99.82	96.61	94.58	99.81	96.52
	3	85.80	84.63	99.82	95.58	96.52	99.82	91.12	96.43	99.81
6	1	99.80	94.34	94.67	99.83	96.96	98.84	99.82	97.13	94.34
	2	94.53	99.81	90.32	97.03	99.84	97.43	97.08	99.82	96.84
	3	94.44	89.54	99.82	98.90	97.39	99.84	94.10	96.75	99.81

Using a fixed parameter a_1 and a dynamic parameter $b_{0,t}$ results in only one parameter adjusting to the dynamics of the system, making the interpretation of the changes easier. From Figure 6 it can be deducted that $b_{0,t}$ is an indicator of how much the cells are competing for the medium. On one hand, if the parameter exhibits an overall low absolute value, as seen in Figure 6a, it indicates that the cells have leftover medium that is not used and will be replaced unnecessarily, which means that resources are wasted. On the other hand, if an overall higher absolute value is attained, as seen in Figure 6b, then the cells do not have enough medium to fulfil their potential growth.

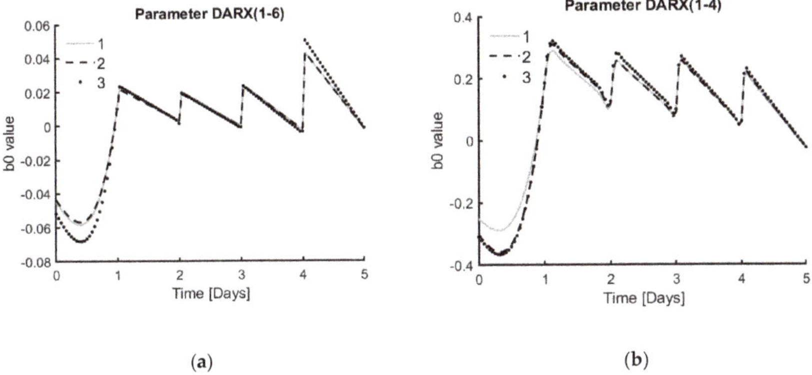

Figure 6. Graphs exhibiting the $b_{0,t}$ values over time. (**a**) $b_{0,t}$ values for triplicates of donor 1 under condition 6, which means 100% medium replacement every 24 h. (**b**) $b_{0,t}$ values for triplicates of donor 1 under condition 4, which means only 10% medium replacement every 24 h.

3.4. Simulation of the Model Predictive Controller

MPC simulations were performed based on the identified prediction models for each type of medium replacement strategy. An example for condition 1 and condition 6 are given in Figure 7, with accumulated lactate produced by the cells as target output and accumulated amount of medium replacement as an input variable.

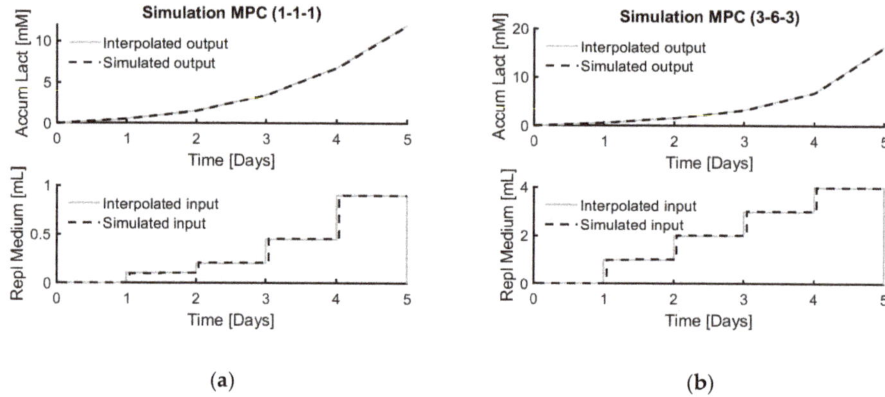

Figure 7. Top graph shows the interpolated output (accumulated lactate (mM)) compared to the simulated output. Bottom graph represents the interpolated input (total replaced medium (mL)) compared to the simulated input. (**a**) model-based predictive control (MPC) simulation applied to donor 1, condition 1 and triplicate 1; (**b**) MPC simulation applied to donor 3, condition 6 and triplicate 3.

The goodness of fit between the controller's input suggestions and output and the experimental data are summarised in Tables 6 and 7.

Table 6. The accuracy of the MPC simulation is measured using NRMSE and multiplied by 100 to be expressed as a percentage, with 100 being a perfect fit. The accuracy of the MPC simulation is represented for the difference in input (accumulated replaced medium (mL)) of the experimental data compared to the input of the simulated data. All NRMSE values calculated for all three donors and all three triplicates are equal for the same condition of medium replacement.

Condition	1	2	3	4	5	6
NRMSE	84.96	86.28	86.18	87.18	87.07	87.03

Table 7. The accuracy of the MPC simulation is measured using NRMSE and multiplied by 100 to be expressed as a percentage, with 100 being a perfect fit. The accuracy of the MPC simulation is represented as the difference between output (accumulated lactate (mM)) of the experimental data compared to the output of the simulated data. The DARX model used to perform the simulation is either the one based on the corresponding experimental data (diagonal values) or on the data of an experimental triplicate.

NRMSE			Donor 1			Donor 2			Donor 3		
			DARX Model Triplicate								
Condition			1	2	3	1	2	3	1	2	3
1	Experimental data triplicate	1	98.76	98.77	98.77	98.73	98.72	98.73	98.65	98.64	98.65
		2	98.76	98.76	98.77	98.73	98.73	98.73	98.64	98.63	98.64
		3	98.82	98.82	98.83	98.68	98.67	98.68	98.67	98.66	98.67
2		1	98.70	98.70	98.71	98.67	98.66	98.66	98.58	98.57	98.59
		2	98.70	98.70	98.71	98.68	98.67	98.67	98.57	98.56	98.58
		3	98.71	98.71	98.72	98.67	98.66	98.66	98.59	98.58	98.60
3		1	98.67	98.67	98.68	98.63	98.62	98.63	98.58	98.56	98.61
		2	98.68	98.68	98.69	98.65	98.64	98.65	98.56	98.54	98.59
		3	98.68	98.68	98.69	98.61	98.61	98.62	98.56	98.54	98.59
4		1	98.77	98.78	98.78	98.83	98.82	98.84	98.67	98.66	98.67
		2	98.76	98.78	98.78	98.85	98.84	98.85	98.67	98.65	98.66
		3	98.77	98.78	98.78	98.82	98.81	98.82	98.67	98.66	98.67
5		1	98.60	98.59	98.60	98.64	98.64	98.64	98.51	98.50	98.51
		2	98.62	98.62	98.63	98.65	98.65	98.65	98.53	98.52	98.53
		3	98.56	98.56	98.57	98.64	98.64	98.64	98.51	98.50	98.51
6		1	98.55	98.55	98.56	98.53	98.53	98.53	98.47	98.47	98.47
		2	98.53	98.53	98.54	98.53	98.53	98.54	98.48	98.48	98.48
		3	98.49	98.50	98.50	98.51	98.51	98.51	98.47	98.47	98.47

4. Discussion

Monitoring and controlling the cell growth is crucial when developing a large-scale reproducible cell culture process. However, there are currently no standardised methods to sample the amount of cells during a cell culture expansion in tissue flasks or hollow fibre bioreactors. Previous studies have therefore investigated the benefits of controlling the environment of the cell culture vessels using standard physicochemical process parameters [15]. In addition, other studies developed potential soft sensors using the metabolic responses of the cells to control the process, mostly glucose concentration [33,34]. This work used this metabolic soft sensor concept by measuring the lactate concentration off-line and used it as an indication of the cell growth, which can otherwise only be measured at the end of the bioprocess.

Choosing the correct control strategy for this framework results in high accuracy between the experimental data and the simulated data. Many different control strategies have been explored in fermentation processes [35], some for mammalian cells [15,17], and a few for human cells [36]. These control strategies are built on either user experience, a process model or historical data [35]. Each strategy has its own benefits and disadvantages. Using an approach based only on user experience has the advantage that it can be quickly applied to a new system without the need for historical data or a process model. However, these approaches, such as probing control [37] or fuzzy control [38], are running behind the action, because they act when the current state is not ideal, without an optimal strategy for the whole process. When there is a large amount of historical data, interesting approaches are artificial neural networks [20,21] or statistical process controls [39]. However, for cell therapy bioprocesses this is mostly not the case, since these data are very process-specific and cannot be extrapolated for different cell types, batch sizes or in autologous applications, which are donor-specific.

Mechanistic mathematical approaches encounter the same difficulty, because their specific sets of kinetic parameters have to be redefined for each specific process, requiring many specific data sets. A mathematical model, for example, one that describes the exponential growth of cells in combination with consumption nutrients and production waste products [36], is useful for the prediction of an average control strategy for that cell type. However, the downside of these mathematical models is that they contain cell-lineage-specific kinetics parameters from literature and should be updated for every stage of that cell lineage, e.g., proliferation or differentiation [40].

In cases where there is a process model available, the preferred choice would be to use model-based predictive control (MPC), because it can deal with non-linear dynamics, unpredictable disturbances and provides insight for the user [35]. Other attempts at controlling bioprocesses using an MPC have been made. One of them consisted of controlling the glucose concentration to maintain more than a certain threshold of 11 mM in a 15 L fed-batch system [34]. To achieve this, they used a non-linear model-based predictive control to adapt the feed rate based on a mechanistic mathematical model which describes the cell growth and metabolism. However, the main problem was the process–model mismatch, which is inherent to the variability of a bioprocess. They also compared an off-line measurement method with 12 h between samples to an on-line spectroscopy technique sampling every six minutes. The problem with on-line glucose methods was a high sample-to-noise ratio. Another study tried to avoid this problem of the high cost and noise of on-line glucose sensors by developing a soft sensor [33]. This soft sensor uses cumulative oxygen transfer rates, calculated using several on-line measured variables. It defines the correlation between the on-line soft sensor and the real glucose concentration by comparing off-line measures of glucose every 24 h to recalculate the correlation.

What was still missing from most current control strategies is the combination of a model predictive control with an adaptive control strategy to avoid the process–model mismatch [34]. Therefore, this paper uses the MPC approach and implements an adaptive prediction model. This allows the model to predict the next input to achieve the desired output based on all previous inputs and outputs, taking into account unpredictable disturbances or inherent batch variability in bioprocesses by updating the model parameters in real-time. The accuracy of the model fit when using the same model over different medium replacement conditions or different donors can even be below 50%. This is represented in Table A1 where the model for donor 2 is fitted on data of donor 3. This points out the variability between donors and realisations. However, the potential of the approach developed in this work is that the model is estimated and adapted in real-time solely using data for that specific realisation/individual, and thus guarantees a personalised approach.

This work also uses the concept of a soft sensor by using another measurable variable (lactate concentration) to estimate a desired critical quality attribute of the bioprocess (cell number). The flexibility of the controller to react to disturbances as well as process variability is shown by successfully applying the controller to three different donors and six different control strategies in triplicate.

The next step for this work is to implement the controller in real-time to the system and re-evaluate the performance of the controller. The prediction model will be updated with every new data point received from the current experimental run. In future experiments, the idea is to start from the known model structure, which was found to be the best representation for that bioprocess. In this case, the model would be a DARX model with a fixed a_1 parameter and a variable b_0 over time. After gathering enough data points, depending on the measuring frequency, this could be one day. The model will be developed based on the parameters defined by the process at hand. After this initial data gathering period, the model and controller will be updated in real-time using only the data from the current experiment. Using only the fixed model structure from previous experiments would lead to better results compared to other modelling techniques, where the parameter values of previous experiments are also used without tuning them based on experiment-specific data. The MPC approach presented in this work, which uses the data of each specific realisation, results in a model that adapts well to the process at hand.

In addition, this MPC model could potentially address a case study based on giving the process just enough medium to reach a certain percentage of the maximum cell number at harvest. This maximum cell number is estimated when supplying the process with 100% medium every day. However, to practically perform such a controlled process, additional knowledge about the system is required, which can be gained by performing follow-up experiments. One strategy to consider for these experiments is to observe the b_0 values of the DARX model, which is continuously re-estimated with every new data point collected from the experiment at hand. Further analysis could lead to finding certain thresholds for this parameter that would result in reaching a predefined percentage of the maximum achievable cell number at harvest.

Another additional path to explore is to correlate the cumulative lactate produced back to the biomass growth, in order to use the measurements as a soft sensor and estimate the amount of biomass at each lactate sample time point. However, the relation between the number of cells and lactate produced can differ not only between cell types, but also between different medium replacement strategies. In cases where cells receive a very low amount of medium, cells could die due to nutrient and growth factor depletion. Another possibility which could lead to a change in the relation between the amount of cells and the amount of lactate produced is a metabolic alteration (a by-product of glycolysis) by the cells when the amount of replaced medium is low [41]. Therefore, it would also be important in additional experiments to assess different quality attributes of the cells to check whether all process parameters are possible or if certain thresholds on medium replacement are required to avoid changing the quality and characteristics of the cells. The quality could be assessed with live/dead analysis, additional measures such as Lactate Dehydrogenase (LDH) and flow cytometry for MSC markers or the trilineage potential, determining the osteogenic, chondrogenic or adipogenic potential.

When these two steps of real time implementation and translation into cell numbers are combined, the controller could potentially solve case studies using an adaptive reference trajectory where a specific number of cells is required by a specific realistic time period using a minimal amount of medium. This approach is also capable of implementing different manipulated and controlled variables, in case new sensor techniques come onto the market.

Finally, instead of using well plates as a way to keep process costs low and experimental time short for a large amount of experiments, we envisage the use of such tools for suspension bioreactors where progenitor cell populations can be scaled-up for clinical production, allowing, at the same time, the capacity for real-time process adaptation [10,40].

5. Conclusions

The model predictive controller developed in this work is a generic algorithm which requires minimal effort to implement different process parameters and different responses of the system. This controller has the potential to be an inexpensive tool to minimise the costs and time of cell expansions in combination with assured product quality by design (QbD) [42]. Using cumulative lactate concentrations as an output measurement of the controller has proven to be useful in this specific bioprocess setting, where high glucose DMEM was used. However, it is important, when applying this method to a different bioprocess, to first assess which output measurement and related process parameter would suit that specific bioprocess.

Six different combinations of medium replacement were tested on three different donors in triplicate in order to model the dynamical response of medium replacement on cell proliferation. This dynamic response is best modelled using a DARX prediction model, resulting in an overall high R^2 of 99.80% ± 0.02% for the DARX model on the same experimental data. The process–model mismatch is also low when applying a model based on experimental data from one triplicate to experimental data from another one of the triplicates. The average fit for the triplicates in DARX models on all the triplicates of experimental data is 96.57% ± 3.26%.

Based on simulations, the model predictive controller designed in this work shows promising results to accurately predict the effect of medium replacement on cell growth. The medium change input suggested by the simulation has a 86.45% ± 0.78% accuracy compared to the real experimental data, whereas the accumulated lactate output has an accuracy of 98.64% ± 0.10% compared to the target experimental data.

The results in this work show that this lactate-based model predictive controller can be applied to different donors as well as different medium-replacement strategies. The parameters are estimated for each individual experimental run, resulting in a high accuracy fit between the simulated data and the experimental data. Using these individualised parameters is the main advantage compared to other control strategies, which are more focused on a suitable prediction for the average bioprocess [14,17].

Author Contributions: Conceptualisation, K.V.B., T.L.; methodology, K.V.B., T.L., A.Y., I.P., A.P.F., J.-M.A.; software, K.V.B., A.P.F., A.Y.; validation, K.V.B.; formal analysis, K.V.B.; investigation, K.V.B., I.P.; resources, I.P., J.-M.A.; data curation, K.V.B.; writing—original draft preparation, K.V.B.; writing—review and editing, K.V.B., A.Y., A.P.F., T.L., I.P. and J.-M.A.; visualisation, K.V.B.; supervision, I.P. and J.-M.A.; project administration, J.-M.A.; funding acquisition, I.P., J.-M.A. All authors have read and agreed to the published version of the manuscript.

Funding: This research was funded by KU Leuven, grant number C24/17/077.

Conflicts of Interest: The authors declare no conflict of interest.

Appendix A

Table A1. Model mismatch example, where the DARX models of donor 2 are applied to experimental data from donor 3. The results are the NRMSE multiplied by 100, between the DARX model and the data.

NRMSE			DARX Model Donor 2 Triplicates		
Condition			**1**	**2**	**3**
1	Experimental data triplicate donor 3	1	54.16	50.70	59.50
		2	61.82	58.50	66.99
		3	52.98	49.47	58.27
2		1	56.75	50.97	58.41
		2	66.54	61.10	68.14
		3	63.83	58.28	65.50
3		1	57.66	56.96	63.13
		2	55.66	54.89	61.14
		3	49.32	48.60	55.05
4		1	51.67	48.30	57.18
		2	63.02	59.82	68.15
		3	57.41	54.13	62.73
5		1	87.32	87.72	90.86
		2	82.54	83.28	86.94
		3	78.60	79.38	83.17
6		1	90.43	87.27	89.94
		2	87.54	84.30	86.98
		3	84.36	81.07	83.82

References

1. European Medicines Agency. Advanced Therapy Medicinal Products. Available online: https://www.ema.europa.eu/en/human-regulatory/overview/advanced-therapy-medicinal-products-overview (accessed on 18 May 2020).
2. Office of Tissues and Advanced Therapies. Approved Cellular and Gene Therapy Products|FDA. 2020. Available online: https://www.fda.gov/vaccines-blood-biologics/cellular-gene-therapy-products/approved-cellular-and-gene-therapy-products (accessed on 17 July 2020).

3. June, C.H.; O'Connor, R.S.; Kawalekar, O.U.; Ghassemi, S.; Milone, M,C. CAR T cell immunotherapy for human cancer. *Science* **2018**, *359*, 1361–1365. [CrossRef] [PubMed]
4. Luyten, F.P.; Vanlauwe, J. Tissue engineering approaches for osteoarthritis. *Bone* **2012**, *51*, 289–296. [CrossRef] [PubMed]
5. Patel, D.B.; Santoro, M.; Born, L.J.; Fisher, J.P.; Jay, S.M. Towards rationally designed biomanufacturing of therapeutic extracellular vesicles: Impact of the bioproduction microenvironment. *Biotechnol. Adv.* **2018**, *36*, 2051–2059. [CrossRef] [PubMed]
6. Narayanan, H.; Luna, M.F.; von Stosch, M.; Cruz Bournazou, M.N.; Polotti, G.; Morbidelli, M.; Butté, A.; Sokolov, M. Bioprocessing in the Digital Age: The Role of Process Models. *Biotechnol. J.* **2020**, *15*, 1900172. [CrossRef]
7. Quanten, S.; De Valck, E.; Cluydts, R.; Aerts, J.M.; Berckmans, D. Individualized and time-variant model for the functional link between thermoregulation and sleep onset. *J. Sleep Res.* **2006**, *15*, 183–198. [CrossRef]
8. Jung, S.; Panchalingam, K.M.; Wuerth, R.D.; Rosenberg, L.; Behie, L.A. Large-scale production of human mesenchymal stem cells for clinical applications. *Biotechnol. Appl. Biochem.* **2012**, *59*, 106–120. [CrossRef]
9. Simaria, A.S.; Hassan, S.; Varadaraju, H.; Rowley, J.; Warren, K.; Vanek, P.; Farid, S.S. Allogeneic cell therapy bioprocess economics and optimization: Single-use cell expansion technologies. *Biotechnol. Bioeng.* **2014**, *111*, 69–83. [CrossRef]
10. Lambrechts, T.; Papantoniou, I.; Rice, B.; Schrooten, J.; Luyten, F.P.; Aerts, J.M. Large-scale progenitor cell expansion for multiple donors in a monitored hollow fibre bioreactor. *Cytotherapy* **2016**, *18*, 1219–1233. [CrossRef]
11. Mehrian, M.; Lambrechts, T.; Marechal, M.; Luyten, F.P.; Papantoniou, I.; Geris, L. Predicting in vitro human mesenchymal stromal cell expansion based on individual donor characteristics using machine learning. *Cytotherapy* **2020**, *22*, 82–90. [CrossRef]
12. Bersenev, A.; Kili, S. Management of 'out of specification' commercial autologous CAR-T cell products. *Insights* **2018**, *4*, 1051–1058. [CrossRef]
13. Camacho, E.F.; Bordons, C. *Robust Model Predictive Control*; Springer: London, England, 2007; ISBN 978-3-540-76241-6.
14. Kaiser, E.; Kutz, J.N.; Brunton, S.L. Sparse identification of nonlinear dynamics for model predictive control in the low-data limit. *Proc. R. Soc. A Math. Phys. Eng. Sci.* **2018**, *474*, 20180335. [CrossRef] [PubMed]
15. Rafiq, Q.A.; Coopman, K.; Nienow, A.W.; Hewitt, C.J. A quantitative approach for understanding small-scale human mesenchymal stem cell culture—Implications for large-scale bioprocess development. *Biotechnol. J.* **2013**, *8*, 459–471. [CrossRef] [PubMed]
16. Engelking, L.R. Introduction to Glycolysis (The Embden-Meyerhoff Pathway (EMP)). *Textb. Vet. Physiol. Chem.* **2015**, *8*, 153–158.
17. Schop, D.; Janssen, F.W.; van Rijn, L.D.; Fernandes, H.; Bloem, R.M.; de Bruijn, J.D.; van Dijkhuizen-Radersma, R. Growth, Metabolism, and Growth Inhibitors of Mesenchymal Stem Cells. *Tissue Eng. Part A* **2009**, *15*, 1877–1886. [CrossRef] [PubMed]
18. Michl, J.; Park, K.C.; Swietach, P. Evidence-based guidelines for controlling pH in mammalian live-cell culture systems. *Commun. Biol.* **2019**, *2*, 1–12. [CrossRef]
19. Park, H.J.; Lyons, J.C.; Song, C.W. Acidic environment causes apoptosis by increasing caspase activity. *Br. J. Cancer* **1999**, *80*, 1892–1897. [CrossRef]
20. Rotin, D.; Tannock, I.F.; Robinson, B. Influence of Hypoxia and an Acidic Environment on the Metabolism and Viability of Cultured Cells: Potential Implications for Cell Death in Tumors. *Cancer Res.* **1986**, *46*, 2821–2826.
21. De Bari, C.; Dell'Accio, F.; Vanlauwe, J.; Eyckmans, J.; Khan, I.M.; Archer, C.W.; Jones, E.A.; McGonagle, D.; Mitsiadis, T.A.; Pitzalis, C.; et al. Mesenchymal multipotency of adult human periosteal cells demonstrated by single-cell lineage analysis. *Arthritis Rheum.* **2006**, *54*, 1209–1221. [CrossRef]
22. Bordons, C.; Camacho, E.F. *Model Predictive Control*; Springe: Berlin, Germany, 1999; ISBN 3-540-76241-8.
23. Rawlings, J.B.; Mayne, D.Q. *Model Predictive Control: Theory and Design*; Nob Hill Publishing, LLC: Madison, WI, USA, 2009.
24. Ljung, L. *System Identification: Theory for the User*; Prentice Hall: Englewood Cliffs, NJ, USA, 1987; ISBN 0138816409.
25. Youssef, A.; Exadaktylos, V.; Berckmans, D.A. Towards real-time control of chicken activity in a ventilated chamber. *Biosyst. Eng.* **2015**, *135*, 31–43. [CrossRef]

26. Bayen, A.M.; Siauw, T. Interpolation. In *An Introduction to MATLAB® Programming and Numerical Methods for Engineers*; Academic Press: Cambridge, MA, USA, 2015; Chapter 14; pp. 211–223.
27. Taylor, C.J.; Pedregal, D.J.; Young, P.C.; Tych, W. Environmental time series analysis and forecasting with the Captain toolbox. *Environ. Model. Softw.* **2007**, *22*, 797–814. [CrossRef]
28. Fernández, A.P.; Youssef, A.; Heeren, C.; Matthys, C.; Aerts, J.M. Real-time model predictive control of human bodyweight based on energy intake. *Appl. Sci.* **2019**, *9*, 2609. [CrossRef]
29. Bemporad, A.; Lawrence Ricker, N.; Morari, M. *Model Predictive Control Toolbox TM Reference How to Contact MathWorks - R2019b*; MathWorks: Natick, MA, USA, 2019; p. 994.
30. Patel, S.D.; Papoutsakis, E.T.; Winter, J.N.; Miller, W.M. The lactate issue revisited: Novel feeding protocols to examine inhibition of cell proliferation and glucose metabolism in hematopoietic cell cultures. *Biotechnol. Prog.* **2000**, *16*, 885–892. [CrossRef] [PubMed]
31. Yuan, J.; Kroemer, G. Alternative cell death mechanisms in development and beyond. *Genes Dev.* **2010**, *24*, 2592–2602. [CrossRef] [PubMed]
32. Yuan, X.; Logan, T.M.; Ma, T. Metabolism in human mesenchymal stromal cells: A missing link between HMSC biomanufacturing and therapy? *Front. Immunol.* **2019**, *10*, 977. [CrossRef]
33. Goldrick, S.; Lee, K.; Spencer, C.; Holmes, W.; Kuiper, M.; Turner, R.; Farid, S.S. On-Line Control of Glucose Concentration in High-Yielding Mammalian Cell Cultures Enabled Through Oxygen Transfer Rate Measurements. *Biotechnol. J.* **2018**, *13*, 1700607. [CrossRef]
34. Craven, S.; Whelan, J.; Glennon, B. Glucose concentration control of a fed-batch mammalian cell bioprocess using a nonlinear model predictive controller. *J. Process Control* **2014**, *24*, 344–357. [CrossRef]
35. Mears, L.; Stocks, S.M.; Sin, G.; Gernaey, K.V. A review of control strategies for manipulating the feed rate in fed-batch fermentation processes. *J. Biotechnol.* **2017**, *245*, 34–46. [CrossRef]
36. Galvanauskas, V.; Simutis, R.; Nath, S.C.; Kino-oka, M. Kinetic modeling of human induced pluripotent stem cell expansion in suspension culture. *Regen. Ther.* **2019**, *12*, 88–93. [CrossRef]
37. Johnsson, O.; Andersson, J.; Lidén, G.; Johnsson, C.; Hägglund, T. Feed rate control in fed-batch fermentations based on frequency content analysis. *Biotechnol. Prog.* **2013**, *29*, 817–824. [CrossRef]
38. Hisbullah, M.; Hussain, A.; Ramachandran, K.B. Design of a fuzzy logic controller for regulating substrate feed to fed-batch fermentation. *Food Bioprod. Process. Trans. Inst. Chem. Eng. Part C* **2003**, *81*, 138–146. [CrossRef]
39. Kourti, T.; Nomikos, P.; MacGregor, J.F. Analysis, monitoring and fault diagnosis of batch processes using multiblock and multiway PLS. *J. Process Control* **1995**, *5*, 277–284. [CrossRef]
40. Kresnowati, M.T.A.P.; Forde, G.M.; Chen, X.D. Model-based analysis and optimization of bioreactor for hematopoietic stem cell cultivation. *Bioprocess Biosyst. Eng.* **2011**, *34*, 81–93. [CrossRef] [PubMed]
41. Selleri, S.; Bifsha, P.; Civini, S.; Pacelli, C.; Dieng, M.M.; Lemieux, W.; Jin, P.; Bazin, R.; Patey, N.; Marincola, F.M.; et al. Human mesenchymal stromal cell-secreted lactate induces M2-macrophage differentiation by metabolic reprogramming. *Oncotarget* **2016**, *7*, 30193–30210. [CrossRef] [PubMed]
42. Food and Drug Administration. ICH Harmonised tripartite guideline. In *Proceedings of the International Conference on Harmonisation of Techncal Requirements for Registration of Pharmaceuticals for Human Use*; ICH: Geneva, Switzerland, 2009; p. Q8(R2).

 *bioengineering*

Article

An Approach towards a GMP Compliant In-Vitro Expansion of Human Adipose Stem Cells for Autologous Therapies

Valentin Jossen [1,*], Francesco Muoio [2], Stefano Panella [2], Yves Harder [3,4], Tiziano Tallone [2] and Regine Eibl [1]

1 Institute of Chemistry and Biotechnology, Zurich University of Applied Sciences, 8820 Wädenswil, Switzerland; regine.eibl@zhaw.ch

2 Foundation for Cardiological Research and Education (FCRE), Cardiocentro Ticino Foundation, 6807 Taverne, Switzerland; francesco.muoio@cardiocentro.org (F.M.); stefano.panella@cardiocentro.org (S.P.); tiziano.tallone@cardiocentro.org (T.T.)

3 Department of Plastic, Reconstructive and Aesthetic Surgery, Ente Ospedaliero Cantonale (EOC), 6900 Lugano, Switzerland; yves.harder@eoc.ch

4 Faculty of Biomedical Sciences, Università della Svizzera Italiana, 6900 Lugano, Switzerland

* Correspondence: jose@zhaw.ch or valentin.jossen@zhaw.ch; Tel.: +41-58-934-5334

Received: 12 June 2020; Accepted: 15 July 2020; Published: 20 July 2020

Abstract: Human Adipose Tissue Stem Cells (hASCs) are a valuable source of cells for clinical applications (e.g., treatment of acute myocardial infarction and inflammatory diseases), especially in the field of regenerative medicine. However, for autologous (patient-specific) and allogeneic (off-the-shelf) hASC-based therapies, in-vitro expansion is necessary prior to the clinical application in order to achieve the required cell numbers. Safe, reproducible and economic in-vitro expansion of hASCs for autologous therapies is more problematic because the cell material changes for each treatment. Moreover, cell material is normally isolated from non-healthy or older patients, which further complicates successful in-vitro expansion. Hence, the goal of this study was to perform cell expansion studies with hASCs isolated from two different patients/donors (i.e., different ages and health statuses) under xeno- and serum-free conditions in static, planar (2D) and dynamically mixed (3D) cultivation systems. Our primary aim was I) to compare donor variability under in-vitro conditions and II) to develop and establish an unstructured, segregated growth model as a proof-of-concept study. Maximum cell densities of between 0.49 and 0.65 × 10^5 hASCs/cm^2 were achieved for both donors in 2D and 3D cultivation systems. Cell growth under static and dynamically mixed conditions was comparable, which demonstrated that hydrodynamic stresses (P/V = 0.63 W/m^3, τ_{nt} = 4.96 × 10^{-3} Pa) acting at N_{s1u} (49 rpm for 10 g/L) did not negatively affect cell growth, even under serum-free conditions. However, donor-dependent differences in the cell size were found, which resulted in significantly different maximum cell densities for each of the two donors. In both cases, stemness was well maintained under static 2D and dynamic 3D conditions, as long as the cells were not hyperconfluent. The optimal point for cell harvesting was identified as between cell densities of 0.41 and 0.56 × 10^5 hASCs/cm^2 (end of exponential growth phase). The growth model delivered reliable predictions for cell growth, substrate consumption and metabolite production in both types of cultivation systems. Therefore, the model can be used as a basis for future investigations in order to develop a robust MC-based hASC production process for autologous therapies.

Keywords: human adipose stem cells (hASCs); serum- and xeno-free conditions; UrSuppe stem cell culture medium; autologous therapy; kinetic growth modeling; segregated and unstructured growth model

1. Introduction

The successful development and application of cell-based therapies has the potential to treat a number of currently incurable diseases and to improve patient care. It is therefore not surprising that many research activities [1,2] are taking place all over the world in the field of regenerative medicine. However, despite the progress in this field, there are a number of challenges that remain before cell-based therapies can be performed more routinely in clinical practice.

Human Adipose Tissue Stem Cells (hASCs) have demonstrated their potential to target a number of currently incurable clinical conditions [2]. This is not surprising since adipose tissue has recently been discovered to be a novel abundant source of adult stem cells, which can be collected by minimally invasive, low risk procedures for the donors/patients and processed by different techniques [3–5]. Moreover, results from recently performed clinical trials have indicated possible applications in the treatment of acute myocardial infarction, stroke and a host of inflammatory and immune disorders [6]. Human ASCs are also gaining increasing interest in plastic and reconstructive surgical procedures, where a trend towards stem cell-based tissue-engineering strategies is evident. However, the majority of these clinical applications require in-vitro expansion of the cells to deliver an effective therapeutic dose. The intention of the in-vitro expansion step is to manufacture a sufficient number of hASCs under Good Manufacturing Practice (GMP) conditions and in a cost-effective manner [7–10]. The processing of hASCs must be performed in accordance with the Directive 2003/94/EC for cell-based medicinal products [11]. In general, hASC-based therapies can be broadly divided into two categories: patient-specific therapies (autologous) and off-the-shelf therapies (allogeneic). From an economic point of view, the allogeneic therapy approach seems to be the most attractive option at present [12–14]. However, a crucial factor for the economic success of allogeneic cell-based therapies in terms of affordability will depend on whether the patient receiving the stem cell therapy will require immunosuppressive medication. A combined treatment with immunosuppressive drugs will significantly increase the overall life cycle cost of the treatment. In contrast, autologous therapies require careful consideration of regulatory challenges as well as the distribution and delivery of a safe and effective cell-based therapeutic. Furthermore, it is crucial to consider how a cell therapy manufacturing process can be developed to consistently manufacture products from multiple patients/donors [15]. Therefore, technical and biological characterizations of different cultivation technologies, different donors and other biological aspects are important and will support the development of descriptive and predictive models in the future. Achieving consistency and reproducibility in the manufacture of medicinal products is a key requirement for regulatory approval [16,17] and can be achieved to some extent by a reduction in process variations. A key aspect in reducing process variation is the elimination of fetal bovine serum (FBS) in the cell culture medium [18]. Various studies have already shown that serum-free cell culture media can be used in combination with stirred bioreactors and Microcarrier (MC) technology [19–22] in order to expand human mesenchymal stem cells (hMSCs). In addition to the MC-based expansion technology, hollow fiber bioreactors are also frequently used for the hMSC expansion, with which total cell densities of up to 10^9 hMSC can be achieved [23,24]. In contrast to the MC-based expansion, hMSC cell growth occurs inside the hollow fibers, which are permanently flown through with cell culture medium. However, cell harvest could be problematic in these systems and must be carefully developed based on the expansion process. Amini et al. [25] developed a static, wicking matrix bioreactor that provides a thing film of medium that drips onto cells on the scaffold. They used this new bioreactor concept successfully for the expansion of hiPSC-derived pancreatic cells for the production of insulin. Thus, such new bioreactor concepts are also interesting for the expansion of hMSCs.

In contrast to traditional planar and static cultivation systems, MCs (typically in the range of 100–300 μm) provide a surface on which the strictly adherent hASCs can grow in stirred and instrumented bioreactors. The MCs consist of different materials (e.g., polystyrene and gelatin), including synthetic/organic or natural polymers that are synthesized with different porosities and topographies. The careful selection and tuning of the MCs and the serum-free cell culture medium is important and has an influence on the success of in-vitro cultivation.

The aim of this proof-of-concept study was to perform cell expansion experiments with hASCs isolated from two different patients/donors (i.e., different age and health status) under xeno- and serum-free conditions in static, planar (2D) and dynamically mixed (3) cultivation systems. In so doing, we (I) compared the donor variability under in-vitro conditions in two different cultivation systems and (II) developed and established an unstructured, segregated growth model for future investigations. However, due to the limited accessibility of the donor/patient material, only two donors were considered in the present study in order to establish first versions of the growth model. Special emphasis was placed on determining growth-related parameters (i.e., parameters for growth rate and metabolic flux) and comparing cell-specific Critical Quality Attributes (CQAs) during the processing of the hASCs under the static 2D and dynamic 3D process conditions. The growth-related parameters were subsequently used to establish a mathematical growth model for donor-dependent cell growth description, substrate consumption and metabolite production under static 2D and dynamic 3D process conditions.

2. Materials and Methods

2.1. Procurement of Subcutaneous Adipose Tissue from Human Donors

The human adipose tissue samples used in this study (n = 2 donors, referred to as 080 and 085) were obtained from tissue excess originating from surgical interventions performed at the Department of Plastic, Reconstructive and Aesthetic Surgery at the Ospedale Regionale di Lugano (Switzerland). All patients who donated their adipose tissue provided written agreement in compliance with the directives of the local Ethics Committee of the Canton of Ticino (Switzerland), which approved the project and its procedures (project reference number: CE 2915).

The cellular sources used in this study originate from subcutaneous adipose tissue harvested from the abdominal region of female patients undergoing autologous breast reconstruction under general anesthesia. Firstly, depending on the position of the deep inferior epigastric artery and its perforating vessels (DIEP-flap), a symmetrical diamond-shaped abdominal flap was dissected between the umbilicus and the pubis. Any excess subcutaneous adipose tissue, not used for breast reconstruction, was packed into two sterile bags to avoid any contamination and was delivered for further processing of the tissue. The adipose tissue samples were stored at room temperature and processed within 24 h [26] to obtain the Stromal Vascular Fraction (SVF).

2.2. Isolation and Establishment of a Serum-Free hASC Culture

The extraction of the SVF from human adipose tissue and the in-vitro expansion and cryopreservation of the isolated hASCs was performed in accordance with the ethical principles outlined in the Declaration of Helsinki and in compliance with the directives of the Ethics Committee of the Canton of Ticino (Switzerland). The isolated tissue samples were firstly separated from the skin tissue, washed in PBS and homogenized in a blender for 10–15 s (100–400 g of fat tissue). After this initial step, the tissue was digested for 45 min at 37 °C with 0.28 Wünsch Unit/mL of Collagenase AB [27] (Worthington Biochemical Corp., Lakewood, NJ, USA). The enzymatic reaction was stopped by the addition of PBS supplemented with 1% human albumin (CSL Behring AG, Bern, Switzerland). After separating the aqueous phase from the lipid phase, the aqueous phase was collected in a new sterile tube. The cells were subsequently centrifuged and filtered to obtain a fresh SVF.

In order to characterize the SVF, the cells were stained with anti-CD34-BV650, anti-CD45-PC7, anti-CD73-FITC (BioLegend, San Diego, CA, USA), anti-CD146-PE, anti-CD36-APC (Miltenyi BioTech, Bergisch Gladbach, Germany), 7-amino-actinomycin D (7-AAD) (Becton Dickinson, Franklin Lakes, NJ, USA) and Syto40 (Life Technologies from Thermo Fisher Scientific, Waltham, MA, USA). All of the antibodies were titrated to optimize the signal–to–noise ratio and used at a specific concentration (further information can be found in "Supplementary Materials Table S2"). After 20 min of incubation, the erythrocytes were lysed with 1 mL of VersaLyse solution (Beckman Coulter Inc., Brea, CA, USA).

A Forward Scatter Time-of-Flight channel was used to select single cell events, Syto40 DNA marker was used to exclude cellular debris and 7-AAD was used to discriminate between dead and living cells. Cells were acquired using a Cytoflex flow cytometer (Beckman Coulter Inc., Brea, CA, USA). The ASC cell population was defined as $CD45^-$, $CD146^-$, $CD36^-$, $CD34^+$ and $CD73^+$.

After characterization, cells were seeded at a density of 30,000 ASCs/cm^2 in fibronectin precoated plates (Corning Inc., New York City, NY, USA) with our chemically defined serum- and xeno-free stem cell culture medium, called UrSuppe. The stem cell culture medium was changed every 2–3 days, always keeping 50% of the conditioned medium, until the cells reached a confluency of 80–90%. For passaging, the cells were detached from the growth surface by incubating them for 2 min at 37 °C in TrypLE Select [28] (Life Technologies from Thermo Fisher Scientific, MA, USA). After discarding the supernatant, the cells were resuspended in UrSuppe and passaged or used for other experimental investigations in this study.

2.3. hASC Growth Characterization under Planar, Static Conditions (2D Monolayer Expansion)

2D growth characterization of previously isolated hASCs was performed in precoated T_{25}-flasks (5 µg/cm^2 r-fibronectin; Sigma Aldrich, St. Louis, MO, USA) with the UrSuppe stem cell culture medium (5 mL). For this purpose, the cryopreserved, patient-derived hASCs (P1, 080-$PDL_{cum.}$ 3.9, 085-$PDL_{cum.}$ 3.7) were thawed and precultured in T_{75}-flasks (10,000 hASCs/cm^2; 37 °C, 5% CO_2, 80% rH) in order to achieve the required cell numbers to inoculate 22 × T_{25}-flasks per donor (P2, 080-$PDL_{cum.}$ 6.3, 085-$PDL_{cum.}$ 6.5, 10,000 cells/cm^2). The hASC growth characteristics were assessed over 11 days by harvesting two T_{25}-flasks per donor (2 mL TrypLE Select at 37 °C, 2 min) every day. The cell density, substrate and metabolite measurements were carried out using a NucleoCounter NC-200 (Chemometec, Allerod, Denmark) and a Cedex Bio (Roche Diagnostics, Rotkreuz, Switzerland), respectively. In addition to standard T_{25}-flasks, T_{25}-flasks equipped with pH and DO sensor spots (PreSens, Regensburg, Germany) were also inoculated in parallel for each donor in order to assess the pH and DO profiles during cell growth (37 °C, 5% CO_2, 80% rH). In each case, partial medium exchanges of 40% and 60% were performed for each donor on days 4 and 8.

2.4. hASC Growth Characterization under Dynamically Mixed Conditions (Microcarrier-Based Expansion)

3D growth characterization was performed for each donor (P3, 080-$PDL_{cum.}$ 11.5, 085-$PDL_{cum.}$ 11.6) using fibronectin-coated polystyrene beads (ProNectin® F-COATED, Pall SoloHill, New York City, NY, USA) in 125 mL disposable Corning spinner flasks (=100 mL UrSuppe). An initial cell density of 15,000 cells/cm^2 (=54,000 cells/mL) and a Microcarrier (MC) concentration of 10 g/L (=1 g, 360 cm^2) were used to inoculate the spinner flasks. The MC concentration of 10 g/L was defined based on previous investigations by Schirmaier et al. [29] and Jossen et al. [1,30]. The cell inoculum was prepared in T_{75}-flasks coated with r-fibronectin (5 µg/cm^2) and with cells from P1 (=080-$PDL_{cum.}$ 3.9, 085-$PDL_{cum.}$ 3.7). Before inoculation, the MCs were prepared and sterilized according to the vendor recommendations one day before usage. After cell inoculation, a static cell attachment phase of 24 h was performed in a cell culture incubator (37 °C, 5% CO_2, 80% rH) to allow the cells to attach to the MC surface. After the static attachment phase, the culture was continuously stirred at 49 rpm. The selected impeller speed, which corresponded to the N_{s1u} criterion for 10 g/L MCs in the 125 mL disposable Corning spinner flask, was defined based on experimental and numerical fluid flow investigations by Kaiser et al. [31] and Jossen et al. [1,30]. The N_{s1u} suspension criterion defines the lower limit of N_{s1} (=N_{js}), meaning that some MC beads are still in contact with the reactor bottom, but none of them were at rest [32]. On day 5, a partial medium exchange of 50% was performed. For this purpose, the impeller was switched off and the MCs were allowed to settle. Fifty percent of the working volume was replaced with fresh preheated UrSuppe stem cell culture medium, and the impeller was restarted. No MC feeds were performed during the cultivations.

Off-line samples were taken daily to measure substrate and metabolite concentrations (Glc, Lac and Amn) with a Cedex Bio (Roche Diagnostics, Rotkreuz, Switzerland). After the cells had been detached from the MC surface by the enzymatic treatment (15 min with TrypLE Select), the hASC cell number was measured using a NucleoCounter NC-200. The measured cell specific values were used to calculate the growth-related parameters as described in Section 2.6. In addition to the cell measurements, 1 mL of the MC-cell suspension was fixed immediately after sampling with a 3% paraformaldehyde solution for 4′,6-diamidin-2-phenyliondol (DAPI) staining.

2.5. Cell Analytics

2.5.1. Flow Cytometric Analysis

Flow cytometric measurements were performed at the end of the growth characterization experiments (10th day of cultivation): 2D monolayer and MC-based expansion. The flow cytometric measurements contained different mixtures of the following antibodies: CD26-FITC, CD73-FITC, CD90-APC, CD105-PE (BioLegend, San Diego, CA, USA), CD36-APC, CD146-PE (Miltenyi Biotec, Bergisch Gladbach, Germany), CD55-BV421 (Becton Dickinson, Franklin Lakes, NJ, USA) and CD54-PE (Thermo Fisher Scientific, Waltham, USA). All of the antibodies were titered in advance in order to improve the signal-to-noise ratio; the final measurements were carried out with 50 ng/test (respective mAbs and Isotype controls). A Zombie Yellow™ Fixable Viability Kit (BioLegend, San Diego, CA, USA) was used to distinguish between live and dead cells after fixation (1% paraformaldehyde in DPBS for 1 min at RT). For the staining procedure, 50,000 cells in 100 µL FACS buffer (PBS supplemented with 1% albumin and 50 ng/µL human immunoglobulin, Privigen Immunogobulin, CSL Behring AG, Bern, Switzerland) were pipetted into a well, gently mixed and subsequently incubated in the dark for 15 min at room temperature. After the incubation step, the samples were diluted with 100 µL FACS buffer. Sample acquisition and analysis were performed using a Cytoflex flow cytometer (Beckam Coulter Inc., Brea, CA, USA) and Kaluza analysis software. The spectral spill-over from the different fluorochromes was assessed by spectral compensation of the individual fluorescence channels. For this purpose, single stained control particles (VersaComp Antibody Capture Bead Kit, Beckman Coulter, Brea, CA, USA) or cells in combination with the different fluorochromes were used. The compensation matrix was automatically calculated using the dedicated software function integrated into the Kaluza analysis software. Flow cytometer functionality, including the control of the optical alignment and fluidics, was verified routinely with fluorospheres (CytoFLEX Daily QC fluorospheres, Beckman Coulter, Brea, CA, USA). Further information about the different antibodies can be found in "Supplementary Materials Table S2".

2.5.2. RT-qPCR Analysis

RT-qPCR measurements were carried out at different times (day 1, day 5 and day 10) after daily harvesting of the hASCs. The different RT-qPCR measurement times represent distinct phases of cell proliferation; day 1: start of cell proliferation, day 5: exponential cell growth and day 10: plateau due to cellular confluence (end of cultivation). For this purpose, RNAs were extracted from the cell pellets or from the MCs covered with cells by using a Nucleospin®RNA kit (Macherey-Nagel, Düren, Germany). The RNA purification process included an on-column digestion step with DNase I and was performed according to the manufacturer instructions. The RNA purity and quantity was assessed with a NanoDrop microvolume spectrophotometer (Thermo Fisher, Waltham, MA, USA) and the total RNA integrity was periodically verified by agarose gel analysis. cDNA was obtained from 900 ng RNA using a GoScript™ Reverse Transcription System (Promega, CA, USA). Detailed information about the protocol can be found in "Supplementary Materials Table S3". RT-qPCR of the *PREF1*, *SOX9*, *WISP1*, *WISP2*, *NOTCH1*, *DLL1*, *CD26*, *CD55*, *CD248*, *CD142*, *ZP521*, *ZFP423*, *PPARG*, *DKK1*, *RUNX2*, *CD34*, *CD36* and *CD146* genes was performed using 20 ng cDNA for each gene of interest and a SsoAdvanced™ Universal SYBR® Green Supermix kit (Biorad, Hercules, CA, USA) in combination with a CFX Connect System for signal detection. An overview of the different primer sequences is shown in Table 1, where ACTB was used as an internal control for all measurements. Each primer pair product was checked for proper amplification using agarose gel electrophoresis and only single sharp bands of the expected size were used for further analysis. The RT-qPCR process was divided into 5 phases: (I) initial denaturation (95 °C, 120 s), (II) cycle denaturation (95 °C, 5 s), (III) cycle annealing and extension (60 °C, 20 s), (IV) final denaturation (95 °C, 5 s) and (V) melting curve (65–95 °C, 18 min), where phases II and III were repeated 40 times in the sequence. The resulting data were analyzed using CFX software in order to evaluate the ΔΔCt values, which were normalized using the value of the housekeeping gene ACTB as the reference gene. The relative fold changes of the analyzed genes are related to the beginning of the culture (day 1).

Table 1. Overview of primer sequences used for RT-qPCR measurements.

Genes	Forward Primer (5′-3′)	Reverse Primer (3′-5′)
ACTB	CTG GAA CGG TGA AGG TGA CA	AAG GGA CTT CCT GTA ACA ATG CA
PREF1	TGA CCA GTG CGT GAC CTC T	GGC AGT CCT TTC CCG AGT A
SOX9	AGC GAA CGC ACA TCA AGA C	CTG TAG GCG ATC TGT TGG GG
WISP1	CGA GGT ACG CAA TAG GAG TGT	GAA GGA CTG GCC GTT GTT GTA G
WISP2	GCG ACC AAC TCC ACG TCT G	TCC CCT TCC CGA TAC AGG C
NOTCH1	TGG ACC AGA TTG GGG AGT TC-3′	GCA CAC TCG TCT GTG TTG AC
DLL1	ACT CCG CGT TCA GCA ACC CCA T	TGG GTT TTC TGT TGC GAG GTC ATC AGG
CD26	AGT GGC ACG GCA ACA CAT T	AGA GCT TCT ATC CCG ATG ACT T
CD55	AGA GTT CTG CAA TCG TAG CTG C	CAC AAC AGT ACC GAC TGG AAA AT
CD248	AGT GTT ATT GTA GCG AGG GAC A	CCT CTG GGA AGC TCG GTC TA
CD142	GGC GCT TCA GGC ACT ACA A	TTG ATT GAC GGG TTT GGG TTC
ZFP521	GGC TGT TCA AAC ACA AGC G	GCA CAT TTA TAT GGC TTG TTG
ZFP423	GAT CAC TGT CAG CAG GAC TT	TGC CTC TTC AAG TAG CTC A
PPARG	TGA CAG CGA CTT GGC AAT ATT TAT T	TTG TAG CAG GTT GTC TTG AAT GTC T
DKK1	ATA GCA CCT TGG ATG GGT ATT CC	CTG ATG ACC GGA GAC AAA CAG
RUNX2	TCA ACG ATC TGA GAT TTG TGG G	GGG GAG GAT TTG TGA AGA CGG
CD34	TGG CTG TCT TGG GCA TCA CTG G	CTG AAT GGC CGT TTC TGG AGG TGG
CD36	TGT GCA AAA TCC ACA GGA AGT G	CCT CAG CGT CCT GGG TTA CA
CD146	AGC TCC GCG TCT ACA AAG C	CTA CAC AGG TAG CGA CCT CC

2.6. Determination of Cell Biological Kinetic Parameters: Growth Dynamics and Metabolic Activity

Based on regular measurements of cell density and substrate/metabolite concentration, growth-dependent parameters were calculated for the planar and MC-based cultivations as follows:

(I) Specific growth rate (μ):

$$\mu = \frac{ln(X_A(t)) - ln(X_A(0))}{\Delta t} \tag{1}$$

where μ is the net specific growth rate (d^{-1}), $X_A(t)$ and $X_A(0)$ are the cell numbers (cells/cm^2) at the end and the beginning of the exponential growth phase, respectively, and t is the time (d).

(II) Doubling time (t_d):

$$t_d = \frac{ln(2)}{\mu} \tag{2}$$

where t_d is the doubling time, *ln(2)* the binary logarithm of 2 and μ the specific cell growth rate.

(III) Population Doubling Level (*PDL*):

$$PDL = \frac{1}{log(2)} \cdot log\left(\frac{X_A(t)}{X_A(0)}\right) \tag{3}$$

where PDL is the number of population doublings, and $X_A(0)$ and $X_A(t)$ are the cell numbers (cells/cm^2) at the beginning and the end of the cultivation, respectively.

(IV) Expansion factor (EF):

$$EF = \frac{X_A(t_{max})}{X_A(t = 1)} \tag{4}$$

where EF is the expansion factor and $X_A(t_{max})$ is the maximum cell number and $X_A(t_{=1})$ is the cell number on day 1 (i.e., after cell attachment phase).

(V) Lactate yield from glucose ($Y_{Lac/Glc}$):

$$Y_{Lac/Glc} = \frac{\Delta Lac}{\Delta Glc} \tag{5}$$

where $Y_{Lac/Glc}$ is the lactate yield from glucose, ΔLac is the lactate production over a specific time period and ΔGlc is the glucose consumption over the same time period (=exponential growth phase).

(VI) Specific metabolic flux (q_{met}):

$$q_{met} = \left(\frac{\mu}{X_A(t)}\right)\left(\frac{C_{met}(t) - C_{met}(0)}{e^{\mu t} - 1}\right) \tag{6}$$

where q_{met} is the net specific metabolite consumption or production rate (for Glc, Lac and Amn), μ is the specific cell growth rate (d^{-1}), $X_A(t)$ is the cell number (cells/cm^2) at the end of the exponential growth phase, $C_{met}(t)$ and $C_{met}(0)$ are the metabolite concentrations (mmol/L) at the end and the beginning of the exponential growth phase, respectively, and t is the time (d).

2.7. Modelling of hASC Growth Kinetics in 2D Culture Systems (T_{25}-Flasks)

Based on the findings from the static, planar growth experiments, an unstructured, segregated, simplistic growth model was developed and used to describe the hASC growth kinetics in the T_{25}-flask cultures. A comparable model approach has already been successfully used by Jossen et al. [1] to simulate the anchorage-dependent growth of hASCs on MCs during serum-reduced (5% FBS) expansion in single-use spinner flasks. The same model approach as that employed by Jossen et al. [1] was also used with only minor modifications to simulate MC-based hASCs growth kinetics in this study. Detailed information about the MC-based growth model can be found in Jossen et al. [1] and in "Supplementary Materials".

The general concept for the growth model and the factors that influence the T_{25}-flask cultures are shown in Figure 1. Since hASC growth is anchorage-dependent, possible formation of spheroids in the suspension was not considered in the model. This simplification was justified since no spheroid formation was observed in any of the 2D cultivations that employed an appropriate surface coating (data not shown). Thus, it can be assumed that cells in suspension do not contribute to an increase in the overall cell number, with cell growth restricted to the planar growth surface. To define the starting conditions, it was assumed that initial cell attachment took place during the cell attachment phase, which can be described by the attachment constant k_{at}. After the cells had attached themselves to the planar growth surface, a short cell adaption phase was considered, before the cells began to proliferate.

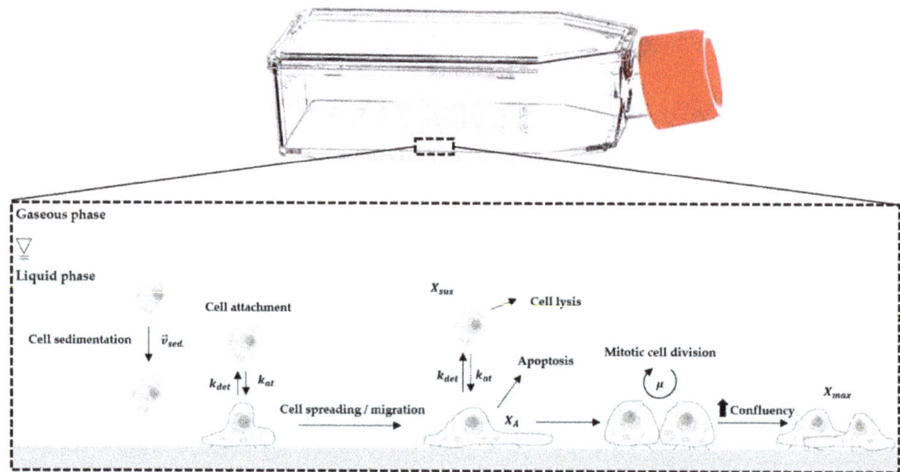

Figure 1. Principle of the growth model and influencing factors.

The cell adaption phase was considered by introducing the coefficient $\alpha(t)$ (see Equation (7)),

$$\alpha(t) = \frac{t^n}{t_l{}^n + t^n} \tag{7}$$

where t_l defines the lag time and the point at which $\alpha(t)$ is half of the maximum. The exponent n in Equation (7) affects the slope of $f(\alpha(t))$. If $n = 1$, $\alpha(t)$ is described by Michaelis–Menten kinetics. Otherwise, a sigmoidal curve is obtained that becomes steeper as n increases. Both variables (t_l and n) were obtained from experimental growth studies using different donor cells.

The specific cell growth rate (μ) was calculated based on Monod-type kinetics. Hence, glucose (Glc), lactate (Lac), ammonium (Amn) and the available growth surface (X_{max}) were considered to be influencing factors (see Equation (8)). However, analysis of microscopic pictures from time lapse microscopic investigations indicated that cell growth restriction based on the maximum available growth surface does not follow a normal Monod-type kinetic (data not shown). This observation can mainly be ascribed to cell migration during cell growth. Thus, the effect of the growth surface restriction term becomes more significant towards the end of the cell growth phase. For this reason, the exponent n was also introduced in Equation (8) as a growth surface restriction term.

$$\mu = \mu_{max} \cdot \left(\frac{Glc}{K_{Glc} + Glc}\right) \cdot \left(\frac{K_{Lac}}{K_{Lac} + Lac}\right) \cdot \left(\frac{K_{Amn}}{K_{Amn} + Amn}\right) \cdot \left(\frac{X_{max}{}^n - X_A{}^n}{X_{max}{}^n}\right) \tag{8}$$

The cell number on the planar growth surface (X_A) increased through mitotic cell division and the attachment of cells from the suspension (see Equation (9)). However, this cell number increase was affected by the detachment of hASCs from the planar growth surface, which was accounted for by the detachment constant ($-k_{det}$).

$$\frac{dX_A}{dt} = \alpha \cdot \mu \cdot X_A + k_{at} \cdot \frac{(X_{max}{}^n - X_A{}^n)}{X_{max}{}^n} \cdot X_{Sus} - k_{det} \cdot X_A \tag{9}$$

Since T_{25}-flasks are static systems, k_{det} is not substantially affected by changing hydrodynamic stresses and can be assumed to be constant. hASC growth in suspension was negligible and therefore changes in cell number were only affected by cell attachment to or detachment from the growth surface (see Equation (10)).

$$\frac{dX_{Sus}}{dt} = k_{det} \cdot X_A - k_{at} \cdot \frac{(X_{max}{}^n - X_A{}^n)}{X_{max}{}^n} \cdot X_{Sus} \tag{10}$$

Glucose consumption was assumed to be limited by the glucose concentration itself (see Equation (11)). In other words, glucose consumption was the result of glucose uptake by the mitotic cells and the maintenance metabolism of the mitotic and non-mitotic cells (X_V). A step response (δ_{Glc}) was implemented in Equation (11) to avoid negative glucose concentrations, even though it was highly improbable that *Glc* was completely consumed during the culture time. This was mainly due to the frequent partial medium exchanges and the theoretically low maximum ratio of cells to *Glc* in the T_{25}-flasks.

$$\frac{dGlc}{dt} = -\frac{1}{Y_{\frac{X}{Glc}}} \cdot \alpha \cdot \mu \cdot \frac{(X_{max}{}^n - X_A{}^n)}{X_{max}{}^n} \cdot X_A - m_{Glc} \cdot \delta_{Glc} \cdot X_V \tag{11}$$

L-glutamine (Gln) consumption was not considered in this model, since metabolic measurements indicated that Gln was not a limiting factor in the T_{25}-flask cultures. Moreover, UltraGlutamine (L-alanyl-L-glutamine) was used in the UrSuppe stem cell culture medium, which had undergone a series of complex degradation steps (i.e., (I): cleavage by extracellular peptidases, (II) degradation of free L-glutamine or absorption into the cells and metabolization). The production of lactate (Lac) and ammonium (Amn) was accounted for by Equations (12) and (13).

$$\frac{dLac}{dt} = q_{Lac} \cdot X_A \cdot \alpha + p_{Lac} \cdot X_V \tag{12}$$

$$\frac{dAmn}{dt} = q_{Amn} \cdot X_A \cdot \alpha + p_{Amn} \cdot X_V \tag{13}$$

All growth-related simulations were performed using MATLAB 2019a (MathWorks Inc., Natick, MA, USA). The set of model equations were solved using the ode15s solver in MATLAB.

3. Results and Discussion

3.1. Isolation of hASCs from Subcutaneous Adipose Tissue (SAT)

The SVF obtained from human subcutaneous adipose tissue is a heterogeneous mixture of cells, which are isolated by enzymatic dissociation. In general, adipocytes represent roughly two-thirds of the total cells extracted and the rest are blood-derived cells, vascular cells, endothelial cells, smooth muscle cells, pericytes, fibroblasts and hASCs. A multiparameter flow cytometric assay was used in the study to determine the absolute cell number for every cell population and to characterize the cells in the SVF. For this purpose, the target hASC population was defined as being positive for $CD34^+$ and $CD73^+$, and negative for $CD36^-$, $CD45^-$ and $CD146^-$ [33,34]. Table 2 provides an overview of the two patients/donors (080 = healthy patient, 085 = post-chemotherapy patient) investigated in this study and the number of living hASCs isolated from their biopsies. For both investigated cases, the number of isolated hASCs was in the range of 5.7–7.7% of the total living cell population and the fraction of hASCs obtained from donor 085 was 35% greater than from donor 80. Based on the information about the number of live hASCs per patient biopsy, the cells were directly seeded in precoated T-flasks (30,000 cells/cm^2) with our xeno- and serum-free UrSuppe stem cell culture medium in order to establish P0.

Table 2. Results obtained from the two different patients.

Donor	Heath Status	Region	Age	Live Cells	Live hASCs	hASCs
(-)	(-)	(-)	(-)	(10^6 cells)	(10^5 cells)	(%)
080	Healthy	Abdomen	46	9.5	5.4	5.7
085	Post-chemotherapy	Abdomen	26	4.8	3.7	7.7

3.2. hASC Growth under Planar, Static Conditions

Figure 2 shows light microscopic pictures of the patient-derived hASCs (a = donor 080, b = donor 085) during the growth characterization study in the T_{25}-flasks. It is clear that cell attachment occurred during the first 4–6 h after cell inoculation. In both cases, ≥98% of the inoculated cells attached to and spread out across the growth surface under the xeno- and serum-free conditions. These cells exhibited typical fibroblast-like or fibroblastoid cell morphology, with minimum and maximum cell diameters in the range of 12–54 μm and 30–291 μm. Interestingly, analysis of the light microscopic pictures showed that the hASCs isolated from donor 085 had a higher average cell area (2480 μm^2, + 24–30%) compared to those from donor 080 (1810 μm^2). Therefore, lower maximum cell densities (= cells/cm^2) can be expected for donor 085, which has an effect on the total cell yield and future process designs. Qualitative analysis of the microscopic pictures showed that the cells began to migrate and proliferate immediately after the cell attachment phase. As expected, cell confluency increased in both cases as a function of the cell number. This resulted in a cell confluency of nearly 80–90% in both cultures after day 5. From day 5 to day 10, the increase in cell confluency slowed down due to the reduced cell proliferation rate, which was caused by the higher frequency of cell contact inhibition. A maximum cell confluency of 95–100% was achieved in both cases by the end of the cultivation studies (=day 10). From a visual point of view, no significant differences in morphology (i.e., shape, granularity) were found between the two donors or during the culture time.

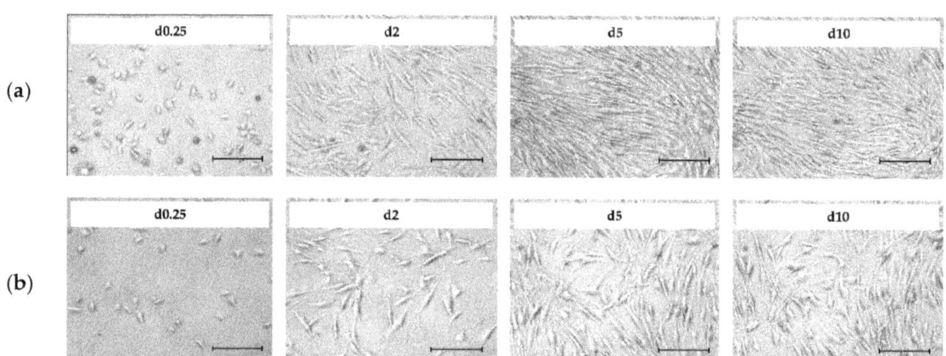

Figure 2. Light microscopic pictures of patient-derived Human Adipose Tissue Stem Cells (hASCs; **a** = donor 080, **b** = donor 085) during cell growth in T_{25}-flasks. Scale bar = 275 μm.

Figure 3a–d shows a quantitative analysis of cell growth and substrate/metabolite concentrations. It is clear that in both cases, the hASCs followed a classical exponential growth curve, which was characterized by four different growth phases: (I) cell adaption phase, (II) exponential growth phase, (III) cell growth restriction phase and (IV) stationary growth phase. A maximum cell density of $0.65 \pm 0.02 \times 10^5$ hASCs/cm^2 (=$3.25 \pm 0.1 \times 10^5$ hASCs/mL) and $0.52 \pm 0.02 \times 10^5$ hASCs/cm^2 (= $2.60 \pm 0.1 \times 10^5$ hASCs/mL) was achieved for donors 080 and 085, respectively. Consequently, the peak cell density of the hASCs from donor 080 was 25% higher than for donor 085. As already mentioned, the differences in the maximum cell densities may be explained by the higher average hASC cell areas from donor 085 compared to donor 080. Interestingly, cell size could be a "consequence" of slow cell growth, in which some cells increase their volume by an increased DNA content and/or other macromolecules [35]. However, these observations need to be further investigated in future studies. During the culture period, cell viability was in both cases always > 95%. The maximum cell densities corresponded to maximum PDLs and EFs in the range of 2.79–3.22 and 7.4–9.9, respectively. Hence, the hASC PDL and EF from donor 080 were 15% and 33% higher, respectively, than for the hASCs from donor 085. Due to the metabolic activity of the cells during the growth phase, glucose was consumed, and lactate and ammonium were produced (Figure 3b,d). In both cultures, the glucose concentration did not drop below 14.04 mmol/L due to the regular partial medium exchanges. Maximum lactate and ammonium concentrations were measured in both cultures in the range of 6.2–6.8 mmol/L and 1.13–1.16 mmol/L. Based on data from Higuera et al. [36] and Schop et al. [37,38] lactate and ammonium did not, however, reach growth-inhibiting concentrations (Lac = 25–35 mM, Amn = 2.5 mM). The online measured pH values (data not shown) agreed well with offline measured data (± 1%) and indicated stable pH values in the region of 7.2–7.3 during the entire cultivation. Cellular respiration caused oxygen to be consumed during cell growth. However, the oxygen supply in the T_{25}-flasks was not a limiting factor and the DO values did not drop below 80% (data not shown). It is clear that by using the developed growth model (see Figure 3a–d lines), the time courses of the cell densities on the MC surface, and the substrate and metabolite concentrations could be well approximated. As indicated by the light microscopic pictures, only a few cells were observed in the supernatant during the static 2D cultivations. Hence, cell density in the supernatant was negligible (simulation results see "Supplementary Material"). Maximum deviations between the measured and simulated cell densities in both cultures were in the range of 8–16%, while slightly higher deviations of up to 21% were found between the measured and simulated substrate/metabolite concentrations. These higher deviations can be explained by (I) uncertainties in substrate/metabolite measurements, (II) error propagation in the calculation of the specific consumption/production rates and (III) slightly different behavior in the real cellular metabolism. Nevertheless, based on growth-dependent parameters, the model successfully describes and predicts cellular growth, substrate consumption and metabolite production in the static 2D cultivation systems.

Table 3 summarizes the calculated growth-dependent parameters for donors 080 and 085. It is clear that with a specific growth rate of 0.52 d^{-1} (t_d = 32 h), the hASCs from donor 080 grew 33% faster than the hASCs from donor 085 (μ_{max} = 0.39 d^{-1}, t_d = 42.7 h). Salzig et al. [20] reported specific growth rates for human bone marrow-derived mesenchymal stem cells (hBM-MSCs) cultivated in a serum-free culture medium in the range of 0.38–0.45 d^{-1} (t_d = 36.9–43.7 h). Comparable specific growth rates (0.31–0.47 d^{-1}) were also reported by Heathman et al. [22] for hBM-MSCs from different donors and over different passages. Therefore, specific hASC growth rates obtained in this study were in a comparable range or even slightly higher (+15%). However, a direct comparison of the specific growth rates is critical since the hMSCs were from different tissue sources and donors and were grown in different serum-free cell culture media. Specific glucose consumption rates ($-q_{Glc}$) were between 1.35 and 1.98 pmol/cell/d and demonstrated that the glucose was more efficiently metabolized by the cells from donor 080. As a result, hASCs from donor 085 produced more lactate for the same equivalent amount of glucose ($Y_{Lac/Glc}$: 1.14 vs. 1.05 mmol/mmol). Ammonium production (0.28–0.32 pmol/cell/d) was comparable in both cultures.

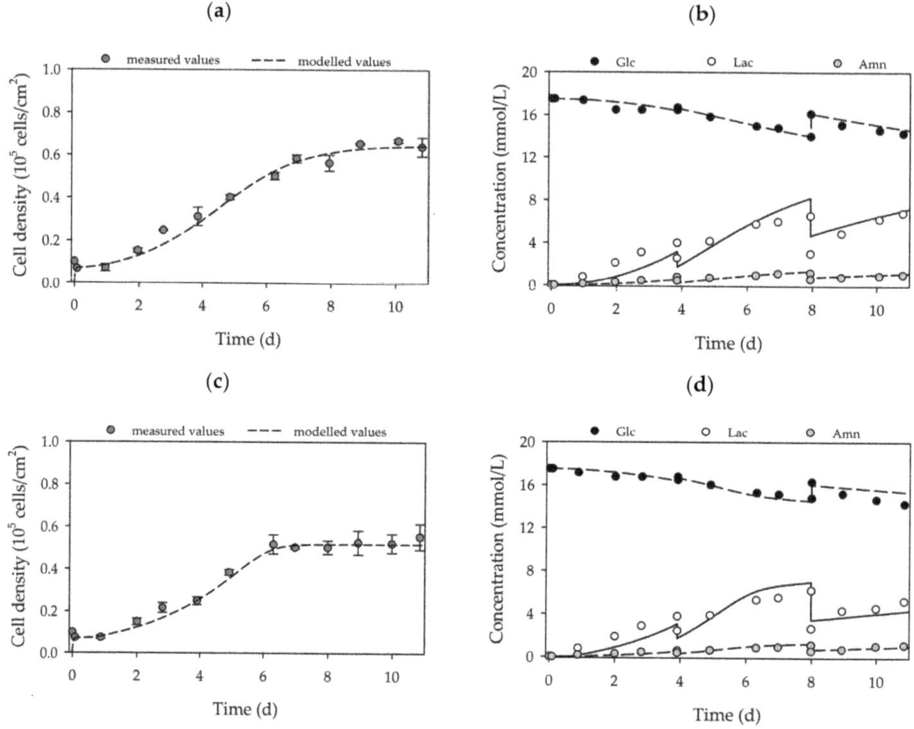

Figure 3. Time-dependent profiles of cell densities (**a**,**c**) and substrate/metabolite concentrations (**b**,**d**) in T_{25}-flasks. Donor 080 (upper row) and 085 (lower row). Partial medium exchanges of 40% and 60% were performed on days 4 and 8, respectively. The symbols represent the experimentally measured values collected from offline measurements. The lines represent the simulated time courses.

Table 3. Overview of the main growth-dependent parameters in the 2D cultivations.

Donor	Xmax $^{(*)}$	PDL $^{(*)}$	EF $^{(**)}$	μ	t_d	$Y_{Lac/Glc}$	qGlc	qLac	qAmn
(-)	(10^5 cells/cm^2)	(-)	(-)	(d^{-1})	(h)	(mmol/mmol)		(pmol/cell/d)	
080	0.65 ± 0.02	3.22 ± 0.04	9.9	0.52	32.0	1.05	1.35	1.41	0.28
085	0.52 ± 0.02	2.79 ± 0.05	7.4	0.39	42.7	1.14	1.98	2.26	0.32

$^{(*)}$ Value was calculated based on the values of the stationary growth phase (mean ± σ_{cells}). (**) Value was calculated based on X_{min} and X_{max}.

3.3. MC-Based hASC Expansion in Single-Use Spinner Flasks

Based on the growth-related parameters obtained from the planar growth characterization studies, cell growth was also characterized for dynamic conditions in MC-based cultivations. Figure 4a,d shows the time-dependent profiles of the cell density and the substrate/metabolite concentrations for hASCs from donors 080 (upper row) and 085 (lower row). After the 24 h cell attachment phase, a cell attachment efficiency of 137% (080) and 118% (085) was achieved. These results indicate that a portion of the cell population had already started to divide within the static cell attachment phase. A peak cell density of 0.61 ± 0.01 × 10^5 hASCs/cm^2 (=2.16 ± 0.04 × 10^5 hASCs/mL) was achieved for donor 080. At 0.49 ± 0.01 × 10^5 hASCs/cm^2 (=1.76 ± 0.04 × 10^5 hASCs/mL), the maximum cell density for donor 085 was again lower (−19%), but nonetheless agreed well with the data from the planar 2D cultivations. In both cases, the maximum cell densities in the MC-based cultivations agreed well with those achieved in the planar cultivation systems. In both cultures, maximum PDLs and EFs were in the range of 1.58–1.72 and 3.2–3.3, respectively. During the cultivation, glucose concentrations decreased to 15.2 mmol/L (080) and 14.3 mmol/L (085), meaning glucose was not a limiting factor in either cultivation. As a result of glucose metabolization, lactate concentration increased to maximum values of 4.5 mmol/L (080) and 5.9 mmol/L (085), and ammonium concentrations remained relatively low during the entire cultivation (080-*Amn* = 0.94 mmol/L, 085-Amn = 0.99 mmol/L). Both lactate and ammonium levels were below critical concentrations [36–38]. From the time-dependent profiles of Figure 4, it can be seen that the growth model can also be used to describe the growth kinetics and substrate/metabolite profiles in MC-based hASC cultivations. The simulated and measured values differed only slightly. Maximum deviations in cell density of ≤7% were found, and deviations in the substrate/metabolite concentrations were only slightly higher (≤15%). Nonetheless, the model can be used in the future to describe hASC growth kinetics in different cultivation settings.

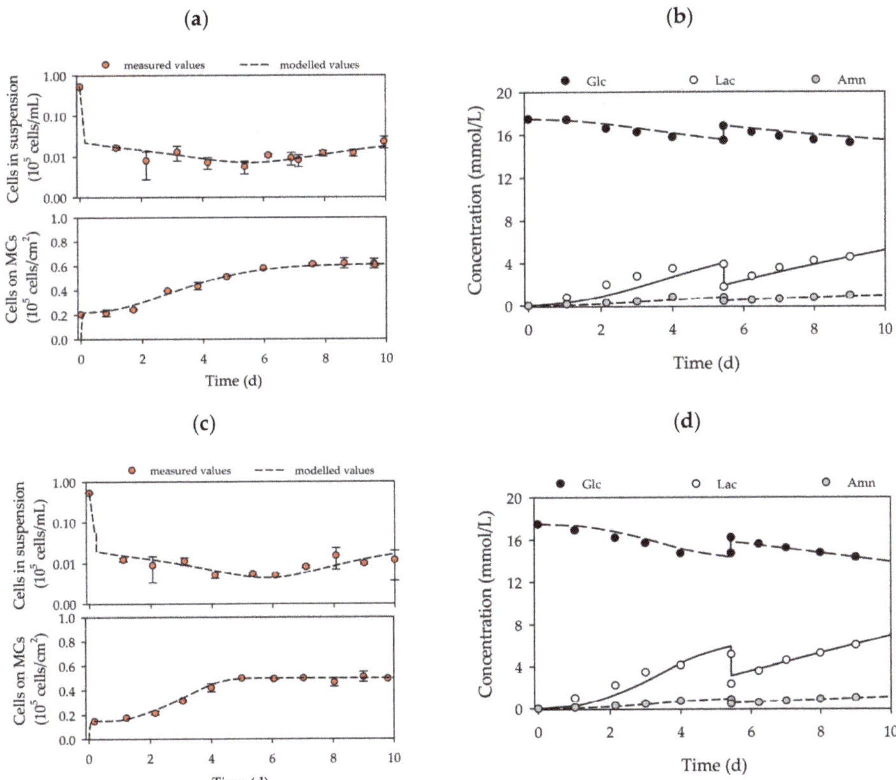

Figure 4. Time-dependent profiles of cell densities (**a**,**c**) and substrate/metabolite concentrations (**b**,**d**) in the Corning spinner flasks. Donor 080 (upper row) and 085 (lower row). A partial medium exchange of 50% was performed on day 5. The symbols represent the experimentally measured values collected by offline measurements. The lines represent the simulated time courses.

Table 4 provides an overview of the main growth-dependent parameters calculated for the two different hASC cultivations. The hASCs from the two different donors grew at comparable specific growth rates of between 0.42 and 0.44 d^{-1} (t_d = 37.8–39.6 h). The calculated specific growth rates were in a comparable range to literature data for MC-based expansions of hBM-MSCs in serum-free cell culture media [20–22,39]. In both cases, glucose was metabolized less efficiently by the hASCs, although $-q_{Glc}$ values (1.34–1.96 pmol/cell/d) were comparable to those in the 2D cultures. Therefore, $Y_{Lac/Glc}$ was in the range of 1.39–1.68 mmol/mmol. The less efficient metabolism of glucose was caused by a higher q_{Lac}, which might be a consequence of the hydrodynamic stresses acting on the cells in dynamically mixed systems. The rates of ammonium production in both cultures (0.26–0.27 pmol/cell/d) were comparable with those in the 2D culture systems.

Table 4. Overview of the main growth-dependent parameters in the Corning spinner flasks.

Donor	Xmax $^{(*)}$	PDL $^{(*)}$	EF $^{(**)}$	μ	t_d	$Y_{Lac/Glc}$	qGlc	qLac	qAmn
(-)	(10^5 cells/cm^2)	(-)	(-)	(d^{-1})	(h)	(mmol/mmol)		(pmol/cell/d)	
080	0.61 ± 0.01	1.58 ± 0.01	3.2	0.44	37.8	1.68	1.34	2.24	0.27
085	0.49 ± 0.01	1.72 ± 0.04	3.3	0.42	39.6	1.39	1.96	2.72	0.26

$^{(*)}$ Value was calculated based on the values of the stationary growth phase (mean ± σ_{cells}). $^{(**)}$ Value was calculated based on X_{min} and X_{max}.

Cell growth in the MC-based cultivations was mainly restricted by the growth surface. Figure 5a,b shows fluorescence microscopic pictures of DAPI-stained hASCs on MCs during cultivation in the spinner flasks. It is clear that on day 1 (after the cell attachment phase) nearly all of the MCs were covered by 2–5 cells. On day 4, some of the MCs were already partially covered with cells and the cells had started to form initial MC-cell-aggregates. By the end of the cultivation, almost all of the MCs were part of a MC-cell-aggregate and only a few MCs were floating around as single beads.

(a)

(b)

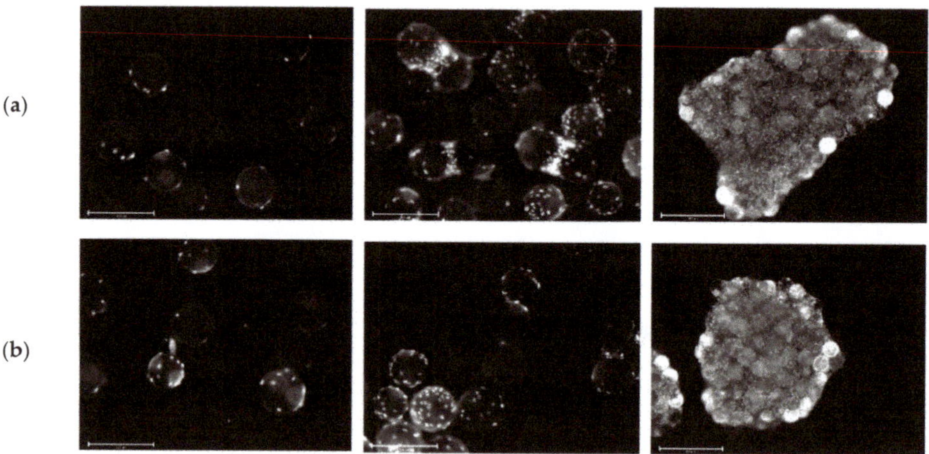

Figure 5. Fluorescence microscopic images during cell growth in Corning spinner flasks. Donors 080 (**a**) and 085 (**b**). DAPI-stained cells on Microcarriers (MCs) on day 1 (left), day 4 (middle) and day 9 (right). Scale bars: 275 μm (left and middle) and 650 μm (right).

Figure 6a,b shows the MC-cell-aggregates at the end of the cultivations (=day 10) and the results of the size distribution analysis of the maximum MC-cell-aggregate diameters. It can be seen that a comparable MC-cell-aggregate size distribution was obtained at the end of both cultivations, with mean MC-cell-aggregate diameters of 1.95 mm (085) and 1.97 mm (080). Minimum and maximum MC-cell-aggregate diameters were measured at 0.8 mm and 5.7 mm for donor 080 and 0.7 mm and 5.0 mm for donor 085. This indicates local volume-weighted hydrodynamic stresses ($\tau_{nt} = 4.96 \times 10^{-3}$ Pa, $\tau_{nn} = 1.15 \times 10^{-3}$ Pa, Jossen et al. [1]) acting on the MC-cell-aggregates controlled their size to some extent. This observation also agreed well with literature findings [40,41]. Furthermore, the results indicated that increased MC-cell-aggregation mainly took place during the stationary growth phase. Therefore, the cell harvest point should also be defined based on MC-cell-aggregate size data.

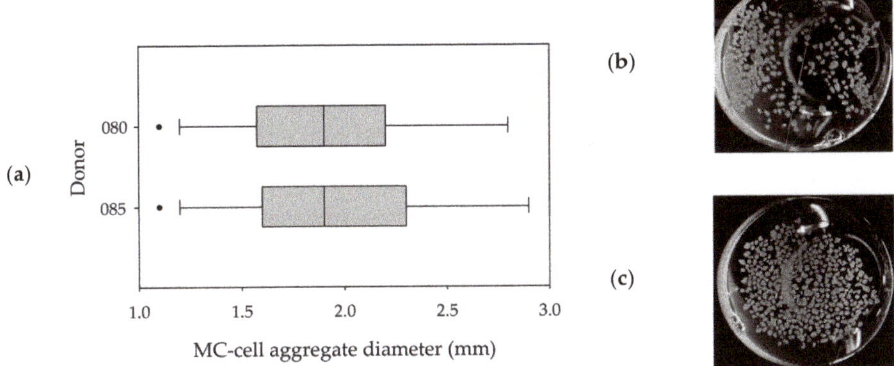

Figure 6. MC-cell-aggregate diameter distributions (**a**) and photographic images (**b**,**c**) of MC-cell aggregates at the end of the cultivations (day 9). Donor 080 (**b**) and 085 (**c**).

3.4. Flow Cytometric Analysis of Standard Markers Expressed by hASCs Cultured in 2D or 3D

Figure 7a,b shows the flow cytometry expression profiles of selected markers analyzed at the end of the cultivation studies (2D vs. 3D). There was no significant difference between the marker expression profile of cells cultured in 2D and those cultured in 3D on MCs. Moreover, no significant differences in the expression profiles were observable between the two donors, which agreed with our expectations. Positive hASC markers ($CD26^+$, $CD54^+$, $CD55^+$, $CD73^+$ and $CD90^+$) were strongly expressed, while negative markers ($CD36^-$ and $CD146^-$) were only weakly expressed. CD105 was the only positive marker that was weakly expressed, which might be caused by the hyperconfluence of the cells after 10 days of cultivation [42–44].

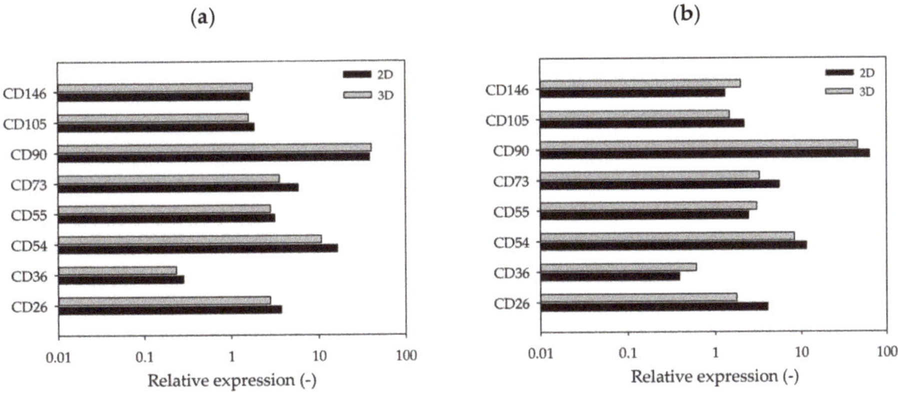

Figure 7. Flow cytometry expression profile of selected markers. hASCs from donor 080 (**a**) and donor 085 (**b**) cultivated in 2D (T_{25}-flasks) or 3D (MC). hASCs were analyzed after harvesting on day 10. Mean fluorescence was calculated based on specific isotype controls (=relative marker expression).

3.5. Monitoring the Expression of Selected Stemness or Cell Differentiation Genes Measured by RT-qPCR

The cells used in this study were extracted from subcutaneous adipose tissue. It is therefore logical to assume that the "default differentiation pathway" of hASCs, in the event of unwanted and uncontrolled spontaneous maturation, is towards adipogenesis. In recent years, several important genes have been discovered that are crucial for the maintenance of stemness and for the differentiation of hASCs [34,45,46]. Further information about the different genes, including their relationships to each other and a short description, can be found in "Supplementary Materials" or in the literature [47–83]. The selected genes can be used as markers in RT-qPCR tests to assess and compare the differentiation status of hASCs expanded in static 2D (see Section 3.2) and/or dynamic 3D conditions (see Section 3.3). To facilitate analysis, the genes were subdivided into three groups:

A. Stemness maintenance genes: *PREF-1, SOX-9, ZFP521, WISP2, NOTCH1 and DLL1*
B. Differentiation regulators/markers: *PPARγ, ZFP423, RUNX2, DKK1, CD34, CD36, CD146* and *WISP1*
C. Lineage hierarchy markers: *CD26, CD55, CD142* and *CD248*

Figure 8 shows the results of the RT-qPCR measurements. In the category, "Stemness Maintenance" (Figure 8a), the cells of both donors on day 5 of the dynamic 3D cultivations had a better profile (high expression of *Pref-1* and *ZFP521*) than the hASCs grown in static 2D conditions. However, after 10 days, the cells were hyperconfluent in both cases. Consequently, the expression of almost all genes decreased [42–44]. Nonetheless, the cells on the MCs performed well and their gene expression pattern was similar to that of those obtained from hASCs grown in static 2D conditions.

In the second category, "Differentiation Regulators/Markers" (Figure 8b), it is clear that for the 3D dynamic conditions, the expression of *PPARγ, RUNX2, DKK1, CD34, CD36* and *WISP1* on day 5 was lower or very similar to the standard 2D set-up. Thus, it can be concluded that the cells retained their "stemness" under the dynamic 3D conditions. As already mentioned, on day 10, the cells were hyperconfluent in both cases. In this situation it is normal that the stemness genes are downregulated, while the differentiation genes are induced [42,44]. However, it is worth noting that the expression of *PPARγ*, the master regulator of adipogenesis, was lower for both donors (080, 085) under dynamic 3D conditions.

In the last category, "Lineage Hierarchy Markers" (Figure 8c), on day 5, the hASCs grown in standard 2D conditions displayed a better gene expression profile than the cells in 3D. However, on day 10, both cell culture systems showed similar satisfactory profiles. It should be noted that the expression of the four genes tested increased during the last days in the 3D spinner system, whereas in the 2D set-up it remained similar between days 5 and 10.

The results of the RT-qPCR measurements clearly showed that the stemness of the hASCs was very well preserved when the cells were grown on xeno-free polystyrene-based MCs with the serum-free UrSuppe stem cell culture medium. The differences in the gene expression profiles were more pronounced on day 5. However, on day 10, due to the very high cell density, most of the stemness genes decreased and most of the differentiation genes increased in both systems, resulting in similar profiles [42–44]. These results clearly demonstrated that for both donors, the optimum point of harvest was on day 5–6 (see Section 3.3). This quality-related observation also agreed very well with the growth-related results and will have an influence on future growth investigations. Hence, higher MC amounts are required to provide the desired cell density, even at lower levels of cell confluency.

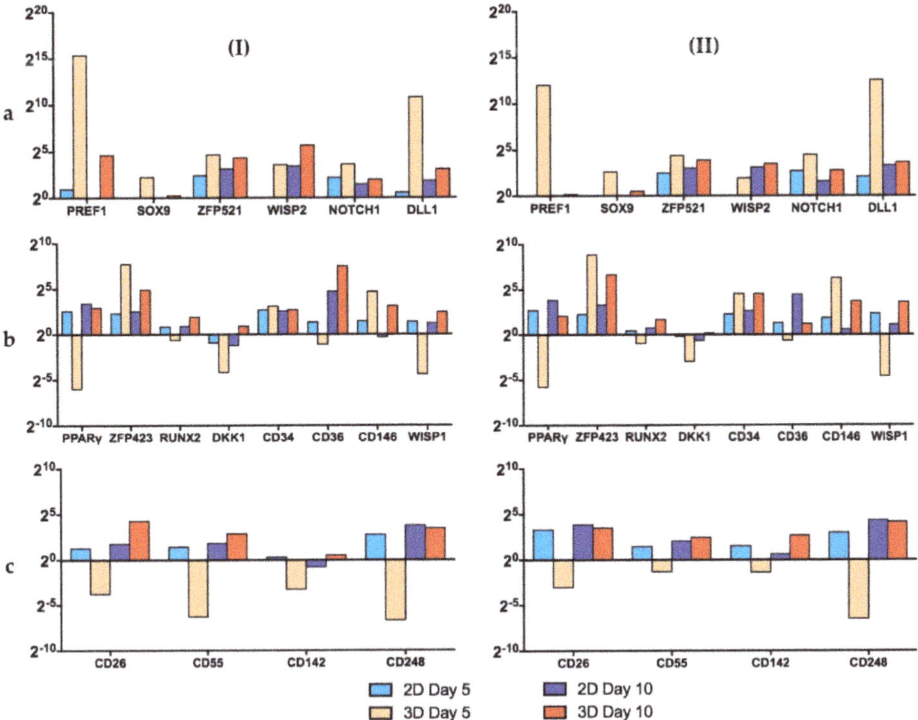

Figure 8. Results of RT-qPCR measurements of donors 080 (I) and 085 (II). The investigated genes were subdivided into 3 groups: (**a**) stemness maintenance genes, (**b**) differentiation regulators/markers and (**c**) lineage hierarchy markers. Data are represented as $2^{(-\Delta\Delta Ct)}$ and related to the beginning of the culture (day 1). A single value for each experimental condition was calculated with this method. This figure is also depicted as "heat maps" and is shown in the "Supplementary Materials" (Figure S3).

4. Conclusions

In this proof-of-concept study, growth- and quality-related investigations were performed under xeno- and serum-free conditions in planar 2D and dynamic 3D cultivation systems with hASCs isolated from two different patients/donors (080 and 085). In order to collect donor-dependent data, two donors of different ages (26 vs. 46 years) and with different health statuses (healthy and post-chemotherapy) were selected for this proof-of-concept study. The hASCs were isolated from the *SVF* under fully serum-free conditions in order to fulfill the regulatory requirements for future hASC manufacturing processes for autologous therapies. The results demonstrated that by using the serum-free UrSuppe stem cell culture medium, hASCs from both donors could be successfully isolated and cultured. The observed tissue frequency of living hASCs was comparable for both donors, although differences in age and health status existed. This is very important for future autologous therapies, as most patients/donors are older or unhealthy. However, further investigations in terms of the biological variation (e.g., gender, broad age range and health conditions) between donors and its effects on successful in-vitro cultivation at different production scales are necessary.

Growth characterization under static 2D conditions revealed differences in the growth performance and the maximum achievable cell densities for the two donors. Due to the higher mean cell areas for donor 085, maximum cell densities were lower, which reduced the overall total cell number per cultivation. The same observations were also performed during the MC-based cultivations. Information about cell morphology (i.e., cell size and area) and the maximum achievable cell density per donor under static 2D conditions is crucial for process scale-up in order to achieve the cell densities required for autologous therapies within the shortest time and number of passages (e.g., issue of cellular senescence). Due to the restricted maximum cell densities for individual donors, high amounts of MCs are necessary in order to increase the total cell number per process step. However, this has an effect on the choice of the stirred bioreactor type and the process conditions. As a result of the shear sensitivity of hASCs, the number of MCs can only be increased up to a certain point, where the occurring hydrodynamic stresses do not negatively affect cell growth or quality. The MC-based expansions clearly showed that the hydrodynamic stresses at N_{s1u} did not significantly affect cell growth or cell quality, even though the stem cell culture medium did not contain FBS. The flow cytometric and RT-qPCR measurements highlighted the maintenance of the stemness during the static 2D and dynamic 3D cultivations of cells from both donors. Surface marker and gene expression profiles under dynamically mixed conditions were comparable for both donors and partially even better than for static 2D conditions. The results also clearly indicate that careful determination of the correct harvest point is important in order to retain stemness. A hyperconfluent culture would increase the total cell density per cultivation but lead to a downregulation of stemness maintenance genes and an upregulation of differentiation marker genes. Thus, optimal cell harvest densities for both donors were determined to be between 0.41 and 0.56 × 10^5 hASCs/cm^2, which were on average 14–22% lower than the maximum cell densities. Consequently, the required MC surface area per cultivation should in the future be defined based on the optimal cell harvest density.

The unstructured, segregated growth model very clearly showed time courses for cell growth, glucose consumption, lactate production and ammonium production that were similar to experimental data from the planar 2D and dynamic 3D cultivations. Maximum deviations for cell density and substrate/metabolite concentrations were in the range of 7–16% and 15–21%, respectively. This means that the descriptiveness power of the model was satisfactory, especially when considering the accuracy of the experimentally measured values. The intensified MC-cell-aggregate formation during the MC-based expansion was not considered in the growth model. Nevertheless, good agreement has been achieved. Therefore, the model can serve as basis for further investigations with hASCs. For this purpose, comprehensive growth studies with hASCs from a larger number of patient/donors (n = 12–20) are planned in stirred, instrumented single-use bioreactors (i.e., BioBLU 0.3c) based on a Design of Experiment approach.

Supplementary Materials: The following are available online at http://www.mdpi.com/2306-5354/7/3/77/s1. Growth model principle and influencing factors in MC-based hASC expansions; Figure S1. Factors that positively or negatively regulate adipogenesis; Figure S2. Time-dependent profiles of cell density in the supernatant of the T25-flasks for donor 080 (a) and 085 (b); Figure S3. Results of RT-qPCR measurements ("heat maps") of donors 080 (I) and 085 (II); Table S1. Parameters used for the kinetic growth model (2D and 3D); Table S2. Detailed information of the antibodies used for the flow cytometric measurements; Table S3. Reverse transcription detailed procedure; Table S4. Overview of measured stemness maintenance genes; Table S5. Overview of measured differentiation regulators/markers; Table S6. Overview of measured lineage hierarchy markers.

Author Contributions: V.J., R.E. and T.T. conceived and designed the experiments; Y.H. selected and recruited the adipose tissue donors, collected the biopsies and prepared the sterile tissue samples for shipment; F.M., S.P. and T.T. processed the human adipose tissue samples and isolated the hASCs; V.J. performed the planar 2D and dynamic 3D cultivation studies; F.M., S.P. and T.T. carried out the cell analytical measurements/analysis (flow cytometry, RT-qPCR); V.J. developed the unstructured, segregated growth model and completed all growth-related simulations and calculations; V.J. wrote the original draft of the paper; F.M., S.P., T.T., R.E. and V.J. wrote, reviewed, and edited the original manuscript. All authors have read and agreed to the published version of the manuscript.

Funding: The cell biology experiments in this study were supported by the Cardiocentro Ticino Foundation (CCTF), and the Foundation for Cardiological Research and Education (FCRE).

Acknowledgments: We are grateful to Dante Moccetti (Director of Cardiocentro Ticino) for his interest and continuous support with the project. We also thank Fabio D'Auria, Fabio Ferri, Giorgio Gaiatto, Luca Perini, and Andreja Petkovic (Logistics Cardiocentro Ticino) for the collection and transport of the biopsies and for the organization of all shipments between Lugano and Wädenswil.

Conflicts of Interest: The authors declare that there is no conflict of interests regarding the publication of this paper.

Latin Symbols

Amn	mmol/L	Ammonium concentration
EF	-	Expansion factor
Glc	mmol/L	Glucose concentration
k_{at}	d^{-1}	Cell attachment constant
k_{det}	d^{-1}	Cell detachment constant
K_{Amn}	mmol/L	Inhibition constant of ammonium
K_{Glc}	mmol/L	Monod constant of glucose
K_{Lac}	mmol/L	Inhibition constant of lactate
Lac	mmol/L	Lactate concentration
N_{s1u}	rpm	Impeller speed at which the MCs are still in contact with the reactor bottom but none of the at rest (lower limit of N_{s1})
PDL	-	Population doubling level
p_{Amn}	mmol/cell/d	Specific ammonium production rate (growth-independent)
p_{Lac}	mmol/cell/d	Specific lactate production rate (growth-independent)
$qAmn$	mmol/cell/d	Specific ammonium production rate (growth-dependent)
$qGlc$	mmol/cell/d	Specific glucose consumption rate
$qLac$	mmol/cell/d	Specific lactate production rate (growth-dependent)
t_d	d	Doubling time of cell population
t_l	d	Lag or cell adaption time
X_A	cells/cm^2	Cell concentration on planar growth surface
X_{max}	cells/cm^2	Maximum cell concentration on planar growth surface
X_{Sus}	cells/mL	Cell concentration in suspension
X_V	cells/cm^2	Cell concentration of viable cells (X_{Sus} + X_A)
$Y_{Lac/Glc}$	mmol/mmol	Lactate yield per glucose equivalent

Greek Symbols

α	-	Cell adaption phase coefficient
β_{MC}	g/L	Microcarrier concentration
δ_{Glc}	-	Step response in glucose balance to avoid negative glucose values (δ_{Glc} = 0 or 1)
μ	1/d	Specific cell growth rate
μ_{max}	1/d	Maximum specific cell growth rate
σ_{cells}	cells/cm^2	Standard deviation of cell density

References

1. Jossen, V.; Eibl, R.; Kraume, M.; Eibl, D. Growth behavior of human adipose tissue-derived stromal/stem cells at small scale: Numerical and experimental investigations. *Bioengineering* **2018**, *5*, 106. [CrossRef] [PubMed]
2. Nordberg, R.C.; Loboa, E.G. Our fat future: Translating adipose stem cell therapy. *Stem Cells Transl. Med.* **2015**, *4*, 974–979. [CrossRef]
3. Duscher, D.; Luan, A.; Rennert, R.; Atashroo, D.; Maan, Z.N.; Brett, E.A.; Whittam, A.; Ho, N.; Lin, M.; Hu, M.S.; et al. Suction assisted liposuction does not impair the regenerative potential of adipose derived stem cells. *J. Transl. Med.* **2016**, *14*, 126. [CrossRef] [PubMed]
4. Oberbauer, E.; Steffenhagen, C.; Wurzer, C.; Gabriel, C.; Redl, H.; Wolbank, S. Enzymatic and non-enzymatic isolation systems for adipose tissue-derived cells: Current state of the art. *Cell Regen.* **2015**, *4*, 7. [CrossRef] [PubMed]
5. De Francesco, F.; Mannucci, S.; Conti, G.; Prè, E.D.; Sbarbati, A.; Riccio, M. A non-enzymatic method to obtain a fat tissue derivative highly enriched in adipose stem cells (ASCs) from human lipoaspirates: Preliminary results. *Int. J. Mol. Sci.* **2018**, *19*, 2061. [CrossRef]
6. Heathman, T.R.; Nienow, A.; McCall, M.; Coopman, K.; Kara, B.; Hewitt, C. The translation of cell-based therapies: Clinical landscape and manufacturing challenges. *Regen. Med.* **2015**, *10*, 49–64. [CrossRef]

7. Izeta, A.; Herrera, C.; Mata, R.; Astori, G.; Giordano, R.; Hernández, C.; Leyva, L.; Arias, S.; Oyonarte, S.; Carmona, G.; et al. Cell-based product classification procedure: What can be done differently to improve decisions on borderline products? *Cytotherapy* **2016**, *18*, 809–815. [CrossRef]
8. Chu, D.-T.; Phuong, T.N.T.; Tien, N.L.B.; Tran, D.-K.; Minh, L.B.; Van Thanh, V.; Anh, P.G.; Pham, V.H.; Nga, V.T. Adipose tissue stem cells for therapy: An update on the progress of isolation, culture, storage, and clinical application. *J. Clin. Med.* **2019**, *8*, 917. [CrossRef]
9. Coecke, S.; Balls, M.; Bowe, G.; Davis, J.; Gstraunthaler, G.; Hartung, T.; Hay, R.; Merten, O.-W.; Price, A.; Schechtman, L.; et al. Guidance on good cell culture practice. *Altern. Lab. Anim.* **2005**, *33*, 261–287. [CrossRef]
10. Sensebé, L.; Gadelorge, M.; Fleury-Cappellesso, S. Production of mesenchymal stromal/stem cells according to good manufacturing practices: A review. *Stem Cell Res. Ther.* **2013**, *4*, 66. [CrossRef] [PubMed]
11. European Union. *Commision Directive 2003/94/EC of 8 October 2003 Laying Down the Principles and Guidelines of Good Manufacturing Practice in Respect of Medicinal Products for Human Use and Investigational Medicinal Products for Human Use*; Europien Union: Luxembourg, 2003; pp. 22–26.
12. Malik, N.N.; Durdy, M.B. Cell therapy landscape. In *Translational Regenerative Medicine*; Elsevier: Amsterdam, The Netherlands, 2015; pp. 87–106. [CrossRef]
13. Simaria, A.S.; Hassan, S.; Varadaraju, H.; Rowley, J.; Warren, K.; Vanek, P.; Farid, S.S. Allogeneic cell therapy bioprocess economics and optimization: Single-use cell expansion technologies. *Biotechnol. Bioeng.* **2014**, *111*, 69–83. [CrossRef] [PubMed]
14. Hassan, S.; Simaria, A.S.; Varadaraju, H.; Gupta, S.; Warren, K.; Farid, S.S. Allogeneic cell therapy bioprocess economics and optimization: Downstream processing decisions. *Regen. Med.* **2015**, *10*, 591–609. [CrossRef]
15. Hourd, P. Regulatory challenges for the manufacture and scale-out of autologous cell therapies. *StemBook* **2014**. [CrossRef] [PubMed]
16. Williams, D.J.; Thomas, R.J.; Hourd, P.C.; Chandra, A.; Ratcliffe, E.; Liu, Y.; Rayment, E.A.; Archer, J.R.; Myers, S. Precision manufacturing for clinical-quality regenerative medicines. *Philos. Trans. R. Soc. A Math. Phys. Eng. Sci.* **2012**, *370*, 3924–3949. [CrossRef] [PubMed]
17. Jossen, V.; Bos, C.V.D.; Eibl, R.; Eibl, D. Manufacturing human mesenchymal stem cells at clinical scale: Process and regulatory challenges. *Appl. Microbiol. Biotechnol.* **2018**, *102*, 3981–3994. [CrossRef]
18. Wappler, J.; Rath, B.; Läufer, T.; Heidenreich, A.; Montzka, K. Eliminating the need of serum testing using low serum culture conditions for human bone marrow-derived mesenchymal stromal cell expansion. *Biomed. Eng. Online* **2013**, *12*, 15. [CrossRef]
19. Carmelo, J.; Fernandes-Platzgummer, A.; Diogo, M.M.; Da Silva, C.L.; Cabral, J.M.S. A xeno-free microcarrier-based stirred culture system for the scalable expansion of human mesenchymal stem/stromal cells isolated from bone marrow and adipose tissue. *Biotechnol. J.* **2015**, *10*, 1235–1247. [CrossRef]
20. Salzig, D.; Leber, J.; Merkewitz, K.; Lange, M.C.; Köster, N.; Czermak, P. Attachment, growth, and detachment of human mesenchymal stem cells in a chemically defined medium. *Stem Cells Int.* **2016**, *2016*, 1–10. [CrossRef]
21. Leber, J.; Barekzai, J.; Blumenstock, M.; Pospisil, B.; Salzig, D.; Czermak, P. Microcarrier choice and bead-to-bead transfer for human mesenchymal stem cells in serum-containing and chemically defined media. *Process Biochem.* **2017**, *59*, 255–265. [CrossRef]
22. Heathman, T.R.; Stolzing, A.; Fabian, C.; Rafiq, Q.A.; Coopman, K.; Nienow, A.; Kara, B.; Hewitt, C. Serum-free process development: Improving the yield and consistency of human mesenchymal stromal cell production. *Cytotherapy* **2015**, *17*, 1524–1535. [CrossRef]
23. Hambor, J. Bioreactor design and bioprocess controls for industrialized cell processing. *BioProcess Int.* **2012**, *10*, 22–33.
24. Hanley, P.J.; Mei, Z.; Durett, A.G.; Cabreira-Harrison, M.D.G.; Klis, M.; Li, W.; Zhao, Y.; Yang, B.; Parsha, K.; Mir, O.; et al. Efficient manufacturing of therapeutic mesenchymal stromal cells with the use of the Quantum Cell Expansion System. *Cytotherapy* **2014**, *16*, 1048–1058. [CrossRef] [PubMed]
25. IAmini, N.; Paluh, J.L.; Xie, Y.; Saxena, V.; Sharfstein, S.T. Insulin production from hiPSC—Derived pancreatic cells in a novel wicking matrix bioreactor. *Biotechnol. Bioeng.* **2020**, *117*, 2247–2261. [CrossRef] [PubMed]
26. Carvalho, P.; Wu, X.; Yu, G.; Dias, I.; Gomes, M.E.; Reis, R.L.; Gimble, J.M. The effect of storage time on adipose-derived stem cell recovery from human lipoaspirates. *Cells Tissues Organs* **2011**, *194*, 494–500. [CrossRef]

27. Carvalho, P.; Gimble, J.M.; Dias, I.; Gomes, M.E.; Reis, R.L. Xenofree enzymatic products for the isolation of human adipose-derived stromal/stem cells. *Tissue Eng. Part C Methods* **2013**, *19*, 473–478. [CrossRef] [PubMed]
28. Carvalho, P.; Wu, X.; Yu, G.; Dietrich, M.; Dias, I.; Gomes, M.E.; Reis, R.L.; Gimble, J.M. Use of animal protein-free products for passaging adherent human adipose-derived stromal/stem cells. *Cytotherapy* **2011**, *13*, 594–597. [CrossRef]
29. Schirmaier, C.; Jossen, V.; Kaiser, S.C.; Jüngerkes, F.; Brill, S.; Safavi-Nab, A.; Siehoff, A.; Bos, C.V.D.; Eibl, D.; Eibl, R. Scale-up of adipose tissue-derived mesenchymal stem cell production in stirred single-use bioreactors under low-serum conditions. *Eng. Life Sci.* **2014**, *14*, 292–303. [CrossRef]
30. Jossen, V.; Schirmer, C.; Sindi, D.M.; Eibl, R.; Kraume, M.; Pörtner, R.; Eibl, D. Theoretical and practical issues that are relevant when scaling up hMSC microcarrier production processes. *Stem Cells Int.* **2016**, *2016*, 1–15. [CrossRef]
31. Kaiser, S.C.; Jossen, V.; Schirmaier, C.; Eibl, D.; Brill, S.; Bos, C.V.D.; Eibl, R. Fluid flow and cell proliferation of mesenchymal adipose-derived stem cells in small-scale, stirred, single-use bioreactors. *Chem. Ing. Tech.* **2013**, *85*, 95–102. [CrossRef]
32. Liepe, F.; Sperling, R.; Jembere Rührwerke, S. *Theoretische Grundlagen, Auslegung und Bewertung*; Fachhochschule: Mittweida, Germany, 1998.
33. Bourin, P.; Bunnell, B.A.; Casteilla, L.; Dominici, M.; Katz, A.J.; March, K.L.; Redl, H.; Rubin, J.P.; Yoshimura, K.; Gimble, J.M. Stromal cells from the adipose tissue-derived stromal vascular fraction and culture expanded adipose tissue-derived stromal/stem cells: A joint statement of the International Federation for Adipose Therapeutics and Science (IFATS) and the International So. *Cytotherapy* **2013**, *15*, 641–648. [CrossRef]
34. Sarantopoulos, C.N.; Banyard, D.; Ziegler, M.E.; Sun, B.; Shaterian, A.; Widgerow, A.D. Elucidating the preadipocyte and its role in adipocyte formation: A comprehensive review. *Stem Cell Rev. Rep.* **2018**, *14*, 27–42. [CrossRef] [PubMed]
35. Guertin, D.A.; Sabatini, D.M. 12—Cell growth. In *The Molecular Basis of Cancer*, 4th ed.; Mendelsohn, J., Gray, J.W., Howley, P.M., Israel, M.A., Eds.; Content Repository Only!: Philadelphia, PA, USA, 2015; pp. 179–190. [CrossRef]
36. Higuera, G.A.; Schop, D.; Janssen, F.; Van Dijkhuizen-Radersma, R.; Van Boxtel, T.; Van Blitterswijk, C.; Van Blitterswijk, C. Quantifying in vitro growth and metabolism kinetics of human mesenchymal stem cells using a mathematical model. *Tissue Eng. Part A* **2009**, *15*, 2653–2663. [CrossRef] [PubMed]
37. Schop, D.; Janssen, F.W.; Van Rijn, L.D.; Fernandes, H.; Bloem, R.M.; De Bruijn, J.D.; Van Dijkhuizen-Radersma, R. Growth, metabolism, and growth inhibitors of mesenchymal stem cells. *Tissue Eng. Part A* **2009**, *15*, 1877–1886. [CrossRef] [PubMed]
38. Schop, D.; Borgart, E.; Janssen, F.W.; Rozemuller, H.; Prins, H.-J.; De Bruijn, J.D.; Van Dijkhuizen-Radersma, R. Expansion of human mesenchymal stromal cells on microcarriers: Growth and metabolism. *J. Tissue Eng. Regen. Med.* **2010**, *4*, 131–140. [CrossRef]
39. Heathman, T.R.; Nienow, A.W.; Rafiq, Q.A.; Coopman, K.; Kara, B.; Hewitt, C. Development of a process control strategy for the serum-free microcarrier expansion of human mesenchymal stem cells towards cost-effective and commercially viable manufacturing. *Biochem. Eng. J.* **2019**, *141*, 200–209. [CrossRef]
40. Ferrari, C.; Balandras, F.; Guedon, E.; Olmos, E.; Chevalot, I.; Marc, A. Limiting cell aggregation during mesenchymal stem cell expansion on microcarriers. *Biotechnol. Prog.* **2012**, *28*, 780–787. [CrossRef]
41. Takahashi, I.; Sato, K.; Mera, H.; Wakitani, S.; Takagi, M. Effects of agitation rate on aggregation during beads-to-beads subcultivation of microcarrier culture of human mesenchymal stem cells. *Cytotechnology* **2016**, *69*, 1–7. [CrossRef]
42. Kim, D.S.; Lee, M.W.; Yoo, K.H.; Lee, T.-H.; Kim, H.J.; Jang, I.K.; Chun, Y.H.; Kim, H.J.; Park, S.J.; Lee, S.H.; et al. Gene expression profiles of human adipose tissue-derived mesenchymal stem cells are modified by cell culture density. *PLoS ONE* **2014**, *9*, e83363. [CrossRef]
43. Kim, D.S.; Lee, M.W.; Ko, Y.J.; Chun, Y.H.; Sung, K.W.; Koo, H.H.; Yoo, K.H. Cell culture density affects the proliferation activity of human adipose tissue stem cells. *Cell Biochem. Funct.* **2016**, *34*, 16–24. [CrossRef]
44. Kim, D.S.; Lee, M.W.; Lee, T.-H.; Sung, K.W.; Koo, H.H.; Yoo, K.H. Cell culture density affects the stemness gene expression of adipose tissue-derived mesenchymal stem cells. *Biomed. Rep.* **2017**, *6*, 300–306. [CrossRef]
45. Cristancho, A.G.; Lazar, M.A. Forming functional fat: A growing understanding of adipocyte differentiation. *Nat. Rev. Mol. Cell Biol.* **2011**, *12*, 722–734. [CrossRef] [PubMed]

46. Lee, M.-J. Hormonal regulation of adipogenesis. In *Comprehensive Physiology*; John Wiley & Sons, Inc.: Hoboken, NJ, USA, 2017; pp. 1151–1195. [CrossRef]
47. Chiarella, E.; Aloisio, A.; Codispoti, B.; Nappo, G.; Scicchitano, S.; Lucchino, V.; Montalcini, Y.; Camarotti, A.; Galasso, O.; Greco, M.; et al. ZNF521 has an inhibitory effect on the adipogenic differentiation of human adipose-derived mesenchymal stem cells. *Stem Cell Rev. Rep.* **2018**, *14*, 901–914. [CrossRef] [PubMed]
48. Ferrand, N.; Béreziat, V.; Moldes, M.; Zaoui, M.; Larsen, A.K.; Sabbah, M. WISP1/CCN4 inhibits adipocyte differentiation through repression of PPARγ activity. *Sci. Rep.* **2017**, *7*, 1749. [CrossRef] [PubMed]
49. Grünberg, J.; Hammarstedt, A.; Hedjazifar, S.; Smith, U. The novel secreted adipokine WNT1-inducible signaling pathway protein 2 (WISP2) is a mesenchymal cell activator of canonical WNT. *J. Biol. Chem.* **2014**, *289*, 6899–6907. [CrossRef] [PubMed]
50. Grünberg, J.R.; Elvin, J.; Paul, A.; Hedjazifar, S.; Hammarstedt, A.; Smith, U. CCN5/W ISP2 and metabolic diseases. *J. Cell Commun. Signal.* **2018**, *12*, 309–318. [CrossRef] [PubMed]
51. Hammarstedt, A.; Hedjazifar, S.; Jenndahl, L.; Gogg, S.; Grünberg, J.; Gustafson, B.; Klimcakova, E.; Stich, V.; Langin, D.; Laakso, M.; et al. WISP2 regulates preadipocyte commitment and PPAR activation by BMP4. *Proc. Natl. Acad. Sci. USA* **2013**, *110*, 2563–2568. [CrossRef]
52. Sul, H.S. Minireview: Pref-1: Role in adipogenesis and mesenchymal cell fate. *Mol. Endocrinol.* **2009**, *23*, 1717–1725. [CrossRef]
53. Hudak, C.S.; Gulyaeva, O.; Wang, Y.; Park, S.-M.; Lee, L.; Kang, C.; Sul, H.S. Pref-1 marks very early mesenchymal precursors required for adipose tissue development and expansion. *Cell Rep.* **2014**, *8*, 678–687. [CrossRef]
54. Kang, S.; Akerblad, P.; Kiviranta, R.; Gupta, R.K.; Kajimura, S.; Griffin, M.; Min, J.; Baron, R.; Rosen, E.D. Regulation of Early Adipose Commitment by Zfp521. *PLoS Biol.* **2012**, *10*, e1001433. [CrossRef] [PubMed]
55. Murahovschi, V.; Pivovarova-Ramich, O.; Ilkavets, I.; Dmitrieva, R.; Döcke, S.; Keyhani-Nejad, F.; Osterhoff, M.; Kemper, M.; Hornemann, S.; Klöting, N.; et al. WISP1 is a novel adipokine linked to inflammation in obesity. *Diabetes* **2015**, *64*, 856–866. [CrossRef]
56. Murata, A.; Yoshino, M.; Hikosaka, M.; Okuyama, K.; Zhou, L.; Sakano, S.; Yagita, H.; Hayashi, S.-I. An evolutionary-conserved function of mammalian notch family members as cell adhesion molecules. *PLoS ONE* **2014**, *9*, e108535. [CrossRef] [PubMed]
57. Ross, D.A.; Rao, P.K.; Kadesch, T. Dual roles for the notch target gene hes-1 in the differentiation of 3T3-L1 preadipocytes. *Mol. Cell. Biol.* **2004**, *24*, 3505–3513. [CrossRef] [PubMed]
58. Shan, T.; Liu, J.; Wu, W.; Xu, Z.; Wang, Y. Roles of notch signaling in adipocyte progenitor cells and mature adipocytes. *J. Cell. Physiol.* **2017**, *232*, 1258–1261. [CrossRef] [PubMed]
59. Sparling, D.P.; Yu, J.; Kim, K.; Zhu, C.; Brachs, S.; Birkenfeld, A.L.; Pajvani, U.B. Adipocyte-specific blockade of gamma-secretase, but not inhibition of Notch activity, reduces adipose insulin sensitivity. *Mol. Metab.* **2016**, *5*, 113–121. [CrossRef] [PubMed]
60. Wang, Y.; Sul, H.S. Pref-1 regulates mesenchymal cell commitment and differentiation through Sox9. *Cell Metab.* **2009**, *9*, 287–302. [CrossRef]
61. Brett, E.; Zielins, E.R.; Chin, M.; Januszyk, M.; Blackshear, C.P.; Findlay, M.; Momeni, A.; Gurtner, G.C.; Longaker, M.T.; Wan, D.C. Isolation of CD248-expressing stromal vascular fraction for targeted improvement of wound healing. *Wound Repair Regen.* **2017**, *25*, 414–422. [CrossRef]
62. Chu, A.J. Tissue factor, blood coagulation, and beyond: An overview. *Int. J. Inflam.* **2011**, *2011*, 1–30. [CrossRef]
63. Merrick, D.; Sakers, A.; Irgebay, Z.; Okada, C.; Calvert, C.; Morley, M.; Percec, I.; Seale, P. Identification of a mesenchymal progenitor cell hierarchy in adipose tissue. *Science* **2019**, *364*, eaav2501. [CrossRef]
64. Metzemaekers, M.; Van Damme, J.; Mortier, A.; Proost, P. Regulation of chemokine activity—A focus on the role of dipeptidyl peptidase IV/CD26. *Front. Immunol.* **2016**, *7*. [CrossRef]
65. Mortier, A.; Gouwy, M.; Van Damme, J.; Proost, P.; Struyf, S. CD26/dipeptidylpeptidase IV-chemokine interactions: Double-edged regulation of inflammation and tumor biology. *J. Leukoc. Biol.* **2016**, *99*, 955–969. [CrossRef]
66. Rennert, R.C.; Januszyk, M.; Sorkin, M.; Rodrigues, M.; Maan, Z.N.; Duscher, D.; Whittam, A.J.; Kosaraju, R.; Chung, M.T.; Paik, K.; et al. Microfluidic single-cell transcriptional analysis rationally identifies novel surface marker profiles to enhance cell-based therapies. *Nat. Commun.* **2016**, *7*, 11945. [CrossRef] [PubMed]

67. Schwalie, P.C.; Dong, H.; Zachara, M.; Russeil, J.; Alpern, D.; Akchiche, N.; Caprara, C.; Sun, W.; Schlaudraff, K.-U.; Soldati, G.; et al. A stromal cell population that inhibits adipogenesis in mammalian fat depots. *Nature* **2018**, *559*, 103–108. [CrossRef] [PubMed]
68. Ahmadian, M.; Suh, J.M.; Hah, N.; Liddle, C.; Atkins, A.R.; Downes, M.; Evans, R.M. PPARγ signaling and metabolism: The good, the bad and the future. *Nat. Med.* **2013**, *19*, 557–566. [CrossRef] [PubMed]
69. Barak, Y.; Nelson, M.C.; Ong, E.S.; Jones, Y.Z.; Ruiz-Lozano, P.; Chien, K.R.; Koder, A.; Evans, R.M. PPARγ is required for placental, cardiac, and adipose tissue development. *Mol. Cell.* **1999**, *4*, 585–595. [CrossRef]
70. Christiaens, V.; Van Hul, M.; Lijnen, H.R.; Scroyen, I. CD36 promotes adipocyte differentiation and adipogenesis. *Biochim. Biophys. Acta Gen. Subj.* **2012**, *1820*, 949–956. [CrossRef]
71. Christodoulides, C. The Wnt antagonist Dickkopf-1 and its receptors are coordinately regulated during early human adipogenesis. *J. Cell Sci.* **2006**, *119*, 2613–2620. [CrossRef]
72. Festy, F.; Hoareau, L.; Bes-Houtmann, S.; Péquin, A.-M.; Gonthier, M.-P.; Munstun, A.; Hoarau, J.J.; Césari, M.; Roche, R. Surface protein expression between human adipose tissue-derived stromal cells and mature adipocytes. *Histochem. Cell Biol.* **2005**, *124*, 113–121. [CrossRef]
73. Gao, H.; Volat, F.; Sandhow, L.; Galitzky, J.; Nguyen, T.; Esteve, D.; Åström, G.; Mejhert, N.; LeDoux, S.; Thalamas, C. CD36 is a marker of human adipocyte progenitors with pronounced adipogenic and triglyceride accumulation potential. *Stem Cells* **2017**, *35*, 1799–1814. [CrossRef]
74. Gupta, R.K.; Arany, Z.; Seale, P.; Mepani, R.J.; Ye, L.; Conroe, H.M.; Roby, Y.A.; Kulaga, H.; Reed, R.R.; Spiegelman, B.M. Transcriptional control of preadipocyte determination by Zfp423. *Nature* **2010**, *464*, 619–623. [CrossRef] [PubMed]
75. Gupta, R.K.; Mepani, R.J.; Kleiner, S.; Lo, J.C.; Khandekar, M.J.; Cohen, P.; Frontini, A.; Bhowmick, D.C.; Ye, L.; Cinti, S. Zfp423 expression identifies committed preadipocytes and localizes to adipose endothelial and perivascular cells. *Cell Metab.* **2012**, *15*, 230–239. [CrossRef] [PubMed]
76. Gustafson, B.; Smith, U. The WNT inhibitor dickkopf 1 and bone morphogenetic protein 4 rescue adipogenesis in hypertrophic obesity in humans. *Diabetes* **2012**, *61*, 1217–1224. [CrossRef]
77. Komori, T. Runx2, an inducer of osteoblast and chondrocyte differentiation. *Histochem. Cell Biol.* **2018**, *149*, 313–323. [CrossRef]
78. Leroyer, A.; Blin, M.G.; Bachelier, R.; Bardin, N.; Blot-Chabaud, M.; Dignat-George, F. CD146 (cluster of differentiation 146): An adhesion molecule involved in vessel homeostasis. *Arter. Thromb. Vasc. Biol.* **2019**, *39*, 1026–1033. [CrossRef]
79. Rosen, E.D.; Sarraf, P.; E Troy, A.; Bradwin, G.; Moore, K.; Milstone, D.S.; Spiegelman, B.M.; Mortensen, R.M. PPARγ is required for the differentiation of adipose tissue in vivo and in vitro. *Mol. Cell.* **1999**, *4*, 611–617. [CrossRef]
80. Scherberich, A.; Di Di Maggio, N.; McNagny, K.M. A familiar stranger: CD34 expression and putative functions in SVF cells of adipose tissue. *World J. Stem Cells* **2013**, *5*, 1. [CrossRef] [PubMed]
81. Sidney, L.E.; Branch, M.J.; Dunphy, S.; Dua, H.S.; Hopkinson, A. Concise review: Evidence for CD34 as a common marker for diverse progenitors. *Stem Cells* **2014**, *32*, 1380–1389. [CrossRef] [PubMed]
82. Tontonoz, P.; Hu, E.; Spiegelman, B.M. Stimulation of adipogenesis in fibroblasts by PPARγ2, a lipid-activated transcription factor. *Cell* **1994**, *79*, 1147–1156. [CrossRef]
83. Walmsley, G.G.; Atashroo, D.A.; Maan, Z.N.; Hu, M.S.; Zielins, E.R.; Tsai, J.M.; Duscher, D.; Paik, K.; Tevlin, R.; Marecic, O.; et al. High-throughput screening of surface marker expression on undifferentiated and differentiated human adipose-derived stromal cells. *Tissue Eng. Part A* **2015**, *21*, 2281–2291. [CrossRef]

bioengineering

Article

Human Embryonic-Derived Mesenchymal Progenitor Cells (hES-MP Cells) are Fully Supported in Culture with Human Platelet Lysates

Sandra M. Jonsdottir-Buch [1,2,3], Kristbjorg Gunnarsdottir [1,2] and Olafur E. Sigurjonsson [1,2,3,4,*]

1 The Blood Bank, Landspitali—The National University Hospital of Iceland, Snorrabraut 60, 101 Reykjavik, Iceland; sandra@platome.com (S.M.J.-B.); oes@ru.is (K.G.)
2 Faculty of Medicine, University of Iceland, Vatnsmyrarvegur 16, 101 Reykjavik, Iceland
3 Platome Biotechnology, Alfaskeid 27, 220 Hafnarfjordur, Iceland
4 School of Science and Engineering, University of Reykjavik, Menntavegur 1, 101 Reykjavik, Iceland
* Correspondence: oes@ru.is; Tel.: +354-543-5523 or +354-694-9427

Received: 5 June 2020; Accepted: 19 July 2020; Published: 20 July 2020

Abstract: Human embryonic stem cell-derived mesenchymal progenitor (hES-MP) cells are mesenchymal-like cells, derived from human embryonic stem cells without the aid of feeder cells. They have been suggested as a potential alternative to mesenchymal stromal cells (MSCs) in regenerative medicine due to their mesenchymal-like proliferation and differentiation characteristics. Cells and cell products intended for regenerative medicine in humans should be derived, expanded and differentiated using conditions free of animal-derived products to minimize risk of animal-transmitted disease and immune reactions to foreign proteins. Human platelets are rich in growth factors needed for cell culture and have been used successfully as an animal serum replacement for MSC expansion and differentiation. In this study, we compared the proliferation of hES-MP cells and MSCs; the hES-MP cell growth was sustained for longer than that of MSCs. Growth factors, gene expression, and surface marker expression in hES-MP cells cultured with either human platelet lysate (hPL) or fetal bovine serum (FBS) supplementation were compared, along with differentiation to osteogenic and chondrogenic lineages. Despite some differences between hES-MP cells grown in hPL- and FBS-supplemented media, hPL was found to be a suitable replacement for FBS. In this paper, we demonstrate for the first time that hES-MP cells can be grown using platelet lysates from expired platelet concentrates (hPL).

Keywords: embryonic stem cells; mesenchymal stromal cells; blood platelets; cell culture techniques; progenitor cells

1. Introduction

The field of regenerative medicine has rapidly expanded in recent years. Various types of somatic stem cells, as well as embryonic and induced pluripotent stem cells, have been evaluated for their regenerative abilities [1–4].

Adult human mesenchymal stromal cells (MSCs) have been studied extensively and are considered promising candidates for regenerative therapies due to their differentiation potential and immunomodulatory function [5–7]. Several limitations are associated with MSCs that may hamper their use in regenerative medicine. These include inter-donor variability, variable isolation protocols, which commonly result in a heterogeneous population of cells, reduced proliferation capacity with cell age, and impairment of biological characteristics following long-term in vitro expansion [5,8–11].

Embryonic stem cells (ESCs) have the potential to provide a uniform and unlimited source of cells with efficient long-term proliferation and differentiation potential [3]. However, ESCs can cause teratoma formation, which reduces their applicability [1]. Various protocols are available that describe the derivation of cells with high MSC resemblance from ESCs [12–18], with the aim of combining the regenerative potential of MSCs with the long-term proliferation of ESCs. ESCs are commonly grown on feeder cells, such as mouse fibroblast cells [19]. Animal-free conditions are, however, recommended when deriving and growing cells intended for human use [1,20,21].

Human embryonic stem cell-derived mesenchymal progenitor (hES-MP) cells resemble MSCs and are derived from human ESCs under feeder-free conditions [22]. They demonstrate efficient long-term proliferation and can differentiate towards osteogenic, chondrogenic, and adipogenic lineages [23]. In addition, they do not form teratomas [23]. For those reasons, they have been suggested as a potential off-the-shelf product to replace MSCs [22,24].

In an effort to provide xeno-free conditions for culturing human cells, it has previously been demonstrated that human platelet lysate (hPL) supplementation is suitable as an animal serum alternative for MSC cultures [25–28]. Platelet lysates can be manufactured from expired human platelet concentrates (PCs) obtained directly from an accredited blood bank; this ensures that the platelets used have been obtained from healthy blood donors and have passed appropriate quality control, mandatory for manufacturing and issuing of blood components [26]. Recycling expired PCs reduces biological waste and avoids competition with blood banks for platelet donors, who are already in high demand [29].

Since hES-MP cells resemble MSCs, we hypothesized that hPL might be a good culture supplement choice to ensure the continuation of an animal-free growth environment. The purpose of this study, therefore, was to determine whether hPL is a suitable replacement for fetal bovine serum (FBS) as a cell culture media supplement for the growth and differentiation of hES-MP cells. Indeed, we demonstrated for the first time that hPLs produced from expired PCs are suitable as culture supplements for hES-MP cells. The lysates allowed hES-MP cells to maintain long-term proliferation, mesenchymal surface marker expression, and differentiation towards bone and cartilage.

2. Materials and Methods

2.1. Ethical Statement

The study was approved by the National Bioethics Committee (number: VSN19-189).

2.2. Preparation and Characterization of hPLs

Ten expired PCs were acquired from the Blood Bank (Reykjavík, Iceland) and frozen (−80 °C) within 24 h after their expiry. Seven PCs were collected by apheresis (donor number, dn = 7), and three PCs were prepared from buffy coats (dn = 15). All PCs were obtained from healthy blood donors of the Blood Bank (22 males and 9 females, aged 18 to 64) and had undergone standard quality control.

Each PC underwent three freeze-thaw cycles (from −80 to 37 °C) to promote platelet lysis. The resulting lysate was platelet-depleted by centrifugation at 5000× g for 20 min. After centrifugation, the supernatant was removed and subjected to a second depletion step. Platelet fragments, visible as a pellet after each centrifugation step, were discarded. The supernatant was filtered through a 0.45 µm filter (Millipore, Billerica, MA, USA), and 40 IU/mL of heparin (Leo Pharma A/S, Ballerup, Denmark) was added. The resulting hPL was aliquoted and stored at −20 °C. Five batches of pooled platelet lysates were prepared. Three batches contained lysate from a buffy coat PC and an apheresis PC, while two batches were made from apheresis PCs only.

The human serum albumin concentration of hPL was evaluated with a Human Albumin ELISA Quantitation Kit (Bethyl Laboratories, Montgomery, TX, USA) to assess variability between the five batches prepared above (Figure S1); no significant batch variability was detected.

Growth factors were measured in both hPL (n = 5, for five hPL batches) and FBS (Gibco, Grand Island, NY, USA; n = 3). The concentrations of bone morphogenic protein 2 (BMP-2), basic fibroblast growth factor (bFGF), vascular endothelial growth factor (VEGF), insulin-like growth factor (IGF), and platelet-derived growth factor BB (PDGF-BB) were evaluated with a standard ELISA development kit (PeproTech, Rocky Hill, NJ, USA), and the transforming growth factor beta (TGF-β) concentration was evaluated with a Human TGF-beta1 Quantikine ELISA Kit (R&D Systems, Minneapolis, MN, USA). The concentrations of growth factors found in hPL were compared to the concentrations found in FBS (Figure S2).

2.3. Cell Culture and Proliferation

Human bone marrow-derived MSCs from three human donors were acquired from Lonza (Walkersville, MD, USA), and hES-MP cells (hES-MP002.5) were donated by Takara Bio Europe AB (previously Cellartis AB), Gothenburg, Sweden [22]. The MSCs used here have been previously used and studied by our group [25,26], and they adhere to International Society for Cellular Therapy (ISCT) criteria, as guaranteed by the manufacturer (the MSCs adhere to plastic under standard culture conditions and can be differentiated to adipogenic, chondrogenic, and osteogenic lineages. The ISCT standards regarding cell surface marker expression are followed). The hES-MP cells (from the same batch reported on in [22]) have previously been shown to differentiate to adipogenic, chondrogenic, and osteogenic lineages and to express several markers of MSCs [22], although they do not expand well on plastic and a gelatin layer must be used (as described below). The representative images of hES-MP cell morphology and of hES-MP cells differentiated into adipogenic, chondrogenic, and osteogenic lineages are provided in Figure S3.

MSCs and hES-MP cells were grown in DMEM/F12+ Glutamax medium (Gibco, Grand Island, NY, USA), 1% penicillin/streptomycin (Gibco), and either 10% hPL (hPL–hES-MP and hPL–MSC treatments/cells) or 10% FBS (FBS–hES-MP and FBS–MSC treatments/cells). The decision to use 10% hPL was made after evaluating different hPL concentrations; our group typically uses 10% hPL in studies with bone marrow-derived MSCs, and this concentration has been investigated in other studies [30–32], which matches the typical concentration of FBS used for supplementation.

The culture surface was coated with 0.1% gelatin (Sigma-Aldrich, St. Louis, MO, USA) to allow hES-MP cell attachment. The cells were grown under standard culture conditions (37 °C, 5% CO_2, and 95% humidity). The medium was changed every 2 to 3 days, and cell passaging was performed when the cells reached 80% to 90% growth confluence. The cells were used for experimentation before reaching passage 8, except when evaluating the long-term proliferation and surface marker expression (10 passages).

Cell proliferation was evaluated with a cell population-doubling (PD) assay over 10 passages for MSCs (n = 3) and hES-MP cells (n = 6). The cells were maintained at standard culture conditions, and the cell count was determined at the end of each passage using a Neubauer hemocytometer (Assistant, Munich, Germany). The number of PDs at each passage was then used to find the cumulative PDs (CPDs) using Equations (1) and (2), as previously described [28,32]:

$$PDs = \frac{\log_{10}(N_H) - \log_{10}(N_0)}{\log_{10}(2)}, \tag{1}$$

$$CPDs = \sum_{i=1}^{n} PDs_i, \tag{2}$$

where N_0 and N_H represent the number of cells seeded and the number of cells harvested (at the end of the expansion period), respectively, and n represents the number of passages.

2.4. Surface Marker Expression

The surface marker expression of hES-MP cells was evaluated with flow cytometry after expansion in either 10% hPL (hPL–hES-MP) or 10% FBS (FBS–hES-MP) for 4, 6, and 10 passages. Cells were stained with CD10, CD13, CD29, CD44, CD45, CD73, CD105, CD184, and HLA-DR antibodies (BD Biosciences, San Jose, CA, USA) according to manufacturer's instructions and then fixed with 0.5%

paraformaldehyde (Sigma-Aldrich) in phosphate-buffered saline (PBS, Gibco). Appropriate isotype-matched controls were used as negative controls. Samples were measured with a FACSCalibur flow cytometer equipped with an argon ion laser (BD Biosciences) and subsequently analyzed using Cellquest Pro software (BD Biosciences). Cells were gated according to their side scatter and forward scatter profiles.

Note that at the time of this study, we were unable to test for CD90. However, in subsequent work (unpublished data), the same batches of hES-MP cells were confirmed to be positive for CD90.

2.5. Osteogenic Differentiation

The hES-MP cells were differentiated towards the osteogenic lineage following two passages in media supplemented with either FBS (FBS–hES-MP) or hPL (hPL–hES-MP). Three thousand hES-MP/cm^2 were seeded and then cultured in hMSC Osteogenic Differentiation Medium BulletKit® (Lonza) for 28 days under standard culture conditions. The differentiation media was fully changed every 2 to 3 days. The cells were harvested after 0, 7, 14, and 28 days of osteogenic differentiation for gene expression analysis, the analysis of alkaline phosphatase (ALP) activity, and the quantification of mineralization with alizarin red staining. The ALP activity was tested with an alkaline phosphatase activity assay (Sigma-Aldrich) and normalized to sample protein concentration using a BCA assay (Pierce Biotechnology, Rockford, IL, USA), as previously described [26]. Mineralization was visualized with alizarin red staining and then quantified after dissolving the bound dye with cetyl-pyridinium chloride (Sigma-Aldrich). The cultures were fixed with 10% formaldehyde (Sigma-Aldrich) and then washed with distilled H_2O prior to being stained with alizarin red for 20 min. After staining, the cultures were washed four times with dH_2O and allowed to dry. Before quantification, the cultures were rehydrated with dH_2O overnight, and then the bound alizarin red dye was dissolved with 10% cetyl-pyridinium chloride for 15 min. Subsequently, the Optical Density (OD) at 562 nm was measured with a photospectrometer (Thermo Scientific, Vantaa, Finland).

2.6. Chondrogenic Differentiation

Chondrogenic differentiation was performed in pellet cultures. First, 2.5×10^5 hES-MP cells were seeded in hMSC Chondrogenic Differentiation Media (BulletKit®, Lonza) in 1.5 mL microtubes (Sarstedt, Nümbrech, Germany). The tubes were then centrifuged at 150× *g* for 5 min to generate a cell pellet. The tube lids were then punctured with a 2 mm sterile needle (Misawa, Tokyo, Japan) to facilitate gas exchange. The medium was fully changed every 2 to 3 days. Chondrogenic pellets were harvested after 0, 7, 14, 28, and 35 days for both gene expression analysis and the evaluation of glycosaminoglycan (GAG) concentration. For GAG analysis, the pellets were washed with PBS three times and then digested for 3 h at 65 °C in a papain extraction reagent containing 0.1 M sodium acetate (Sigma-Aldrich), 0.01 M Na_2EDTA (Carl Roth GmbH + Co. KG, Karlsruhe, Germany), 0.005 M cystein HCl (Sigma), and 8 µL of crystallized papain (Sigma-Aldrich). After papain digestion, the GAG concentration was evaluated with a Blyscan assay (Biocolor, Carrickfergus, UK) following the manufacturer's instructions. The pellets were stained with hematoxylin and eosin staining and Masson's trichrome staining after 28 days of differentiation.

2.7. Gene Expression Analysis

Osteogenic samples for gene expression analysis were harvested with trypsinization after 0, 7, 14, 21, and 28 days of differentiation, put in TRIzol® Reagent (Thermo Scientific) and stored at −80 °C until RNA isolation was performed, which was done within two months. Chondrocytic pellets were harvested after 0, 7, 14, 28, and 35 days of differentiation, washed once in PBS, put in TRIzol® Reagent and stored at −80 °C until RNA isolation within two months. Prior to RNA isolation, samples were homogenized in TRIzol® Reagent using the gentleMACS™ Dissociator (Miltenyi Biotec, Bergisch Gladbach, Germany). RNA isolation was performed according to the TRIzol® Reagent manufacturer's instructions. After RNA isolation, RNA clean-up was performed using an RNeasy Plus Mini Kit

(Qiagen, Hamburg, Germany), following the manufacturer's instructions. Reverse transcription was performed using a GeneAmp® RNA PCR kit (Applied Biosystems, Foster City, CA, USA), according to the manufacturer's instructions. Real-time qPCR was performed on all samples with runt-related transcription factor 2 (*RUNX2*, Hs00231692_m1), secreted phosphoprotein 1 (*SPP1*, Hs00167093_m1), and SRY-box 9 (*SOX9*, Hs01001343_g1) as genes of interest. The two most stable reference genes were carefully chosen after the screening of several reference genes. TATA box-binding protein (*TBP*, Hs00427620_m1) and *YWHAZ* (Hs03044281_g1) were found to be most suitable and were used for the normalization of expression in all samples. TaqMan® Gene Expression Assays (Applied Biosystems) were used in all experiments.

2.8. Statistics

GraphPad® Prism version 5 (GraphPad Software, La Jolla, CA, USA) and Microsoft Excel 2013 were used to analyze all data except gene expression data. Two-way ANOVA was used to analyze statistical significance, followed by a Bonferroni correction as a multiple comparison correction.

Relative gene expression was evaluated using REST-384 © version 2, followed by a pairwise fixed reallocation randomization test to determine statistical significance [33]. Two reference genes, *TBP* and *YWHAZ*, were used for normalization in each sample and relative values presented. Three independent donors of MSCs were used in all experiments. Experiments with hES-MP cells were performed in triplicate. Data are presented as mean ± standard error. A p-value of ≤0.05 was considered statistically significant.

3. Results

3.1. hES-MP Cells Growm in the hPL-Supplemented Medium Demonstrate Consistently High Proliferation Rates over Long Culture Periods

Proliferation was evaluated by counting PDs of MSCs and hES-MP cells at each passage for 10 passages (80 days) (Figure 1). The mean PDs for hES-MP cells were 2.63 ± 0.36 and 2.46 ± 0.374 PDs for FBS and hPL, respectively, whereas the mean PDs for MSCs were 1.69 ± 0.283 PDs for FBS and 0.974 ± 1.32 PDs for hPL. No significant differences in proliferation were noted based on the type of media supplement used. hES-MP cells and MSCs exhibited similar growth kinetics until passage 7 (P7), after which MSC proliferation continuously declined while hES-MP cell proliferation continued at a comparable rate to earlier passages (Figure 1). The difference in proliferation between cell types became significant at P7 ($p \leq 0.05$) and remained significant at every passage thereafter, with the greatest difference observed at P10 (13.7 ± 0.89 CPDs; $p \leq 0.001$).

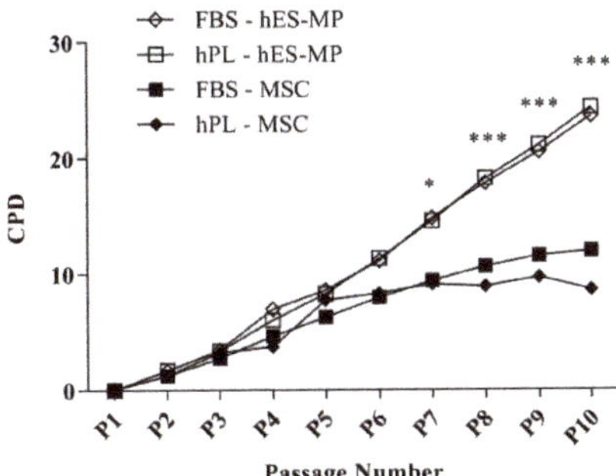

Figure 1. Long-term proliferation of human embryonic stem cell-derived mesenchymal progenitor (hES-MP) cells and mesenchymal stromal cells (MSCs). The MSCs experienced overall lower cumulative population doublings (CPDs) than the hES-MP cells. Differences in cell proliferation were not linked to the type of cell culture supplement used (human platelet lysate (hPL) or fetal bovine serum (FBS)). (* $p \leq 0.05$; *** $p \leq 0.001$).

3.2. hPL–hES-MP and FBS–hES-MP Cells Express CD10, CD13, and CD105 Differently

The surface marker expression in hES-MP cells was evaluated with flow cytometry after expansion in either FBS- or hPL-supplemented media (FBS–hES-MP or hPL–hES-MP, respectively) after P4, P6, and P10. The hES-MP cells strongly expressed CD29, CD44, and CD73, while CD45, CD184, and HLA-DR were expressed only at very low levels (Figure S4). The FBS–hES-MP cells did not express CD10 at P4, P6, and P10 (16.52 ± 0.29, 17.11 ± 0.48, and 15.72 ± 0.20 geometric mean fluorescence intensity (gMFI), respectively) (Figure 2A). The hPL–hES-MP cells, on the other hand, did express CD10. The CD10 expression in the hPL–hES-MP cells was strongest at P6 ($p \leq 0.05$; n = 3), with a fluorescence intensity of 114.6 ± 8.54 gMFI. At P4 and P10, the CD10 expression values were 71.6 ± 0.71 and 61.3 ± 8.74 gMFI, respectively (Figure 2A). Both the FBS–hES-MP and hPL–hES-MP cells were CD13-positive (Figure 2B). However, the CD13 expression remained constant in the FBS–hES-MP cells but increased in the hPL–hES-MP cells over time ($p \leq 0.01$ between P4 and P6; $p \leq 0.05$ between P6 and P10, n = 3). The CD13 expression was significantly stronger in the hPL–hES-MP cells than in the FBS–hES-MP cells at both P6 and P10 (322.8 ± 4.60 and 491.0 ± 56.71 gMFI, $p \leq 0.001$, n = 3). The hES-MP cells expressed CD105 in all cultures, but the expression was significantly lower in the hPL–hES-MP cells than in the FBS–hES-MP cells at P4 ($p \leq 0.01$, n = 3) and P10 ($p \leq 0.001$, n = 3), with the difference in expression being 165.9 ± 10.9 and 228.9 ± 10.0 gMFI, respectively (Figure 2C).

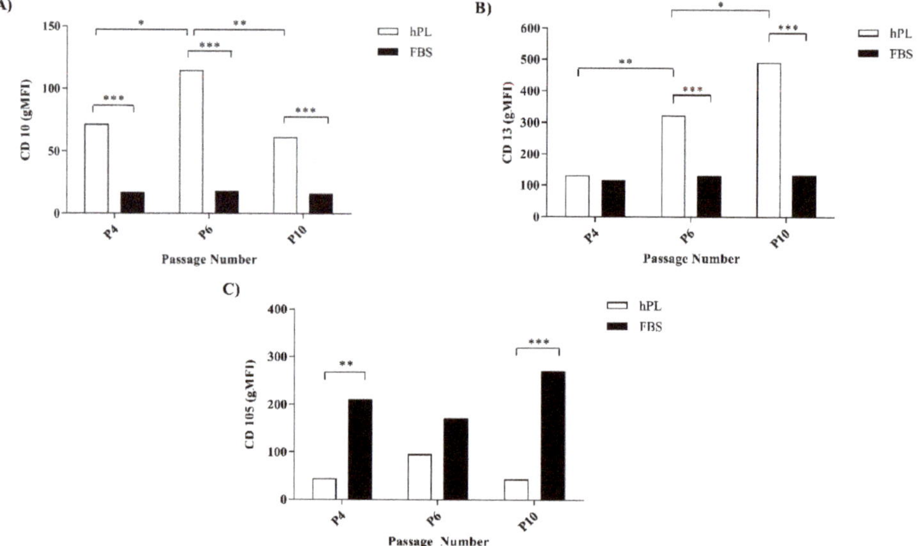

Figure 2. CD10 (**A**), CD13 (**B**), and CD105 (**C**) expression on the hES-MP cell surfaces over 10 passages. The CD10 expression was detected in the hPL–hES-MP cells at all passages, but not in the FBS–hES-MP cells. The CD13 expression was detected in all cultures with the hPL–hES-MP cells, demonstrating time-dependent increases, while the FBS–hES-MP cells showed constant expression. The CD105 expression was significantly lower in the hPL–hES-MP cells than in the FBS–hES-MP cells at all timepoints. (gMFI = geometric mean fluorescence intensity; * $p \leq 0.05$; ** $p \leq 0.01$; *** $p \leq 0.001$).

3.3. Alkaline Phosphatase and Glycosaminoglycans Levels Follow Defined Patterns during Differentiation after Culture in Media Supplemented with either hPL or FBS

The hPL supplementation of cells prior to differentiation was not found to affect the activity of ALP during osteogenic differentiation or the GAG levels during chondrogenesis (Figure 3). Following growth under standard conditions, the FBS–hES-MP and hPL–hES-MP cells were induced towards osteogenic and chondrogenic differentiation after receiving the same differentiation treatment. ALP activity during osteogenic differentiation was evaluated after 0 (control), 7, 14, and 28 days of differentiation (Figure 3A). In the FBS–hES-MP cells, the ALP activity increased from 2.46 ± 0.74 to 39.0 ± 6.10 nmol(p-nitrophenol)/min/mg protein, while in the hPL–hES-MP cells it increased from 5.64 ± 1.21 to 40.62 ± 5.24 nmol(p-nitrophenol)/min/mg protein. The GAG concentration was evaluated as ng of GAG per chondrocytic pellet (Figure 3B). The GAG concentration increased between days 7 and 14, from 249.6 ± 133.1 to 783.4 ± 423.9 ng/pellet in the FBS–hES-MP cells and from 41.9 ± 26.5 to 658.1 ± 359.1 ng/pellet in the hPL–hES-MP cells. After day 14, the GAG concentration gradually decreased. The histology further revealed hypertrophic cells and decrease in cell number at the pellet core by day 28 (Figure S5). No statistically significant differences were observed in ALP activities or GAG concentrations between cultures based on the choice of culture supplement used during expansion.

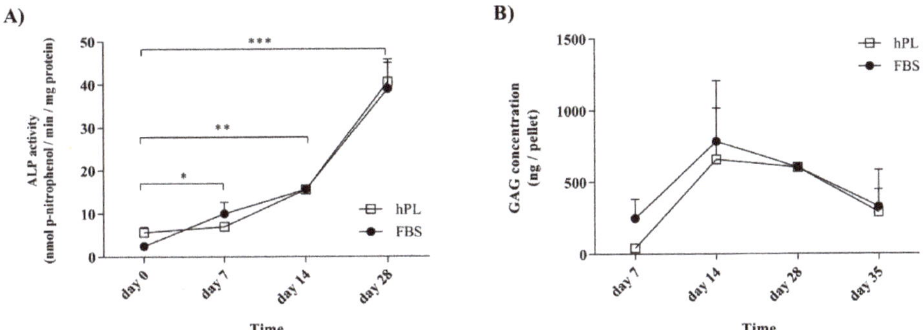

Figure 3. Lineage-specific bioactivity during osteogenic and chondrogenic differentiation. (**A**) During osteogenic differentiation, the alkaline phosphatase (ALP) activity significantly increased in a time-dependent manner over four weeks. (**B**) An initial increase was seen in GAG levels in chondrogenic cultures, with a subsequent decrease after two weeks of differentiation as chondrogenesis advanced. (* $p \leq 0.05$; ** $p \leq 0.01$; *** $p \leq 0.001$).

3.4. hPL–hES-MP Cells Mineralize Later than FBS–hES-MP Cells during Osteogenic Differentiation

The level of mineralization was evaluated after 0 (control), 7, 14, 21, and 28 days of osteogenic differentiation (Figure 4). An increase in mineralization was detected after day 21 for the FBS–hES-MP cells (1.21 ± 0.15 OD; $p \leq 0.001$) and day 28 for the hPL–hES-MP cells (0.96 ± 0.12 OD; $p \leq 0.001$). Mineralization increased significantly between days 21 and 28 in all cultures, with an increase of 0.79 ± 0.12 OD ($p \leq 0.01$) for the FBS–hES-MP cells and 0.90 ± 0.12 OD ($p \leq 0.001$) for the hPL–hES-MP cells. Mineralization was significantly greater in the FBS–hES-MP cells than in the hPL–hES-MP cells at days 21 and 28, with the observed difference being 1.15 ± 0.07 OD ($p \leq 0.001$) and 1.04 ± 0.11 OD ($p \leq 0.001$) on days 21 and 28, respectively (Figure 4).

3.5. Lineage-Associated Gene Expression is Observed during Differentiation following Culturing in the hPL-Supplemented Medium

The expression of *RUNX2* and *SPP1* was evaluated after 7, 14, 21, and 28 days of osteogenic differentiation, and *SOX9* was evaluated after 7, 14, 28, and 35 days of chondrogenic differentiation (Figure 5). The expression was compared to those of undifferentiated FBS–hES-MP and hPL–hES-MP cells. The hPL–hES-MP cells had the significantly higher expression of both *RUNX2* and *SPP1* during the differentiation phase than the FBS–hES-MP cells. The *RUNX2* expression in both FBS–hES-MP and hPL–hES-MP osteoblasts followed a similar pattern but was consistently higher in cells grown with hPL (Figure 5A). The differences ranged from 0.37 ± 0.04 fold ($p \leq 0.05$; n = 3) to 0.53 ± 0.07 fold expression ($p \leq 0.01$; n = 3). The *SPP1* expression in the FBS–hES-MP osteoblasts also followed a similar pattern to that in the hPL–hES-MP osteoblasts, except at day 21, where the *SPP1* expression increased in the hPL–hES-MP cells but continued to decrease in the FBS–hES-MP cells (Figure 5B).

The expression of *SPP1* was consistently higher in the hPL–hES-MP cells than in the FBS–hES-MP cells, with the greatest difference observed at day 21 (0.59 ± 0.03 fold; $p \leq 0.01$; n = 3).

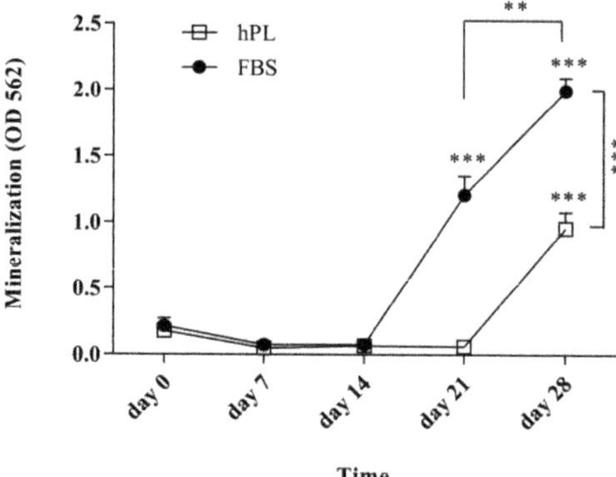

Figure 4. Mineralization during osteogenic differentiation. Significant mineralization was observed after 21 and 28 days for the FBS–hES-MP and hPL–hES-MP cells, respectively. The level of mineralization was significantly higher in the FBS–hES-MP cells than in the hPL–hES-MP cells by the end of 28 days. (** $p \leq 0.01$; *** $p \leq 0.001$; *p*-values directly above data points were compared with that data point on day 0.)

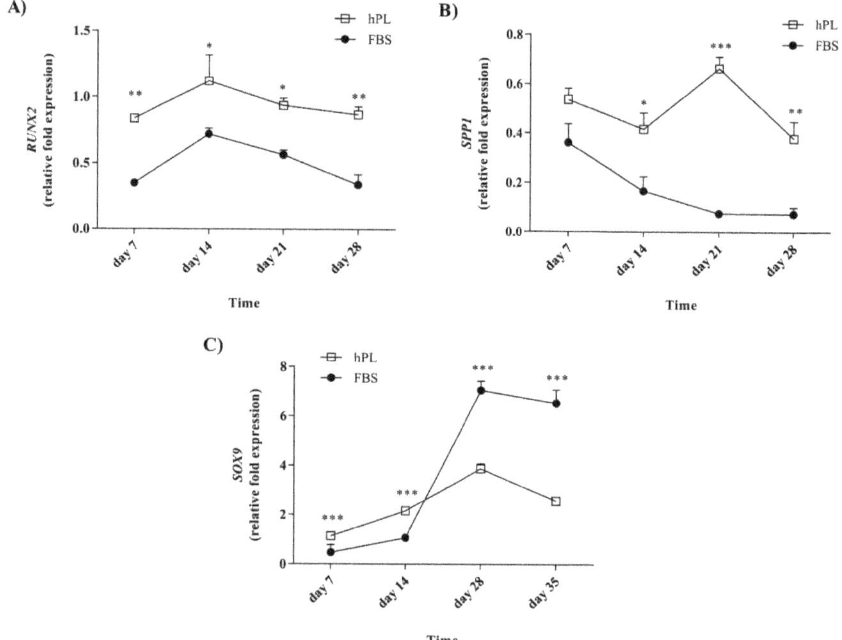

Figure 5. Lineage-specific gene expression during osteogenic and chondrogenic differentiation. During osteogenic differentiation, the hPL–hES-MP cells had the significantly higher expression of both *RUNX2* (**A**) and *SPP1* (**B**) than the FBS–hES-MP cells. (**C**) The *SOX9* expression increased with time in chondrocytes derived from both the FBS–hES-MP and hPL–hES-MP expansion cultures until day 28. The *SOX9* expression was initially higher in the hPL–hES-MP cells, but after day 14 the expression was higher in the FBS–hES-MP cells. (* $p \leq 0.05$; ** $p \leq 0.01$; *** $p \leq 0.001$).

The expression of *SOX9* increased with time in chondrocytes derived from both the FBS–hES-MP and hPL–hES-MP expansion cultures until day 28, after which a slight decrease was observed (Figure 5C). The *SOX9* expression was observed to be higher in hPL chondrocytes at earlier time points, but at day 28, a 4.27 ± 0.04 fold greater increase was observed in the *SOX9* expression in the FBS chondrocytes than in the hPL chondrocytes, after which the expression remained higher in FBS chondrocytes (by 3.18 ± 0.07 fold on day 28) and 3.97 ± 0.08 fold on day 35; $p \leq 0.001$; n = 3).

4. Discussion

hES-MP cells have been reported to have similar characteristics and bioactivity to MSCs [22]. Due to this resemblance, they have been suggested as a potential off-the-shelf product that can be handled and used in an identical manner to MSCs [22,34]. hPL has been widely described as a promising culture supplement for MSCs, but the evidence regarding hES-MP cells has been lacking to date [35]. The goal of the current study was to evaluate, for the first time, how maintaining hES-MP cells in an hPL-supplemented culture medium from expired platelets affects their in vitro growth and differentiation.

When the culture medium was supplemented with hPL, the hES-MP cells demonstrated excellent long-term growth and exhibited the expression of key surface markers previously described for MSCs [36,37]. The evaluation of proliferation over 10 passages revealed that the hES-MP cells exhibited a continuous rate of proliferation for a longer time than the MSCs, with no signs of senescence at the end of the experiment (Figure 1). The MSC growth, on the other hand, began to decline after five passages. No differences in proliferation were found based on whether the hES-MP cells were grown in FBS- or hPL-supplemented media.

The expression of cell surface markers commonly seen on the MSCs was evaluated over 10 passages and compared between treatments (Figure 2). The CD10 expression was detected on the hES-MP cells from hPL-supplemented culture media, but not in FBS-supplemented culture media. The Expression of CD10 on MSCs has been a subject of controversy for a long time, with previous studies reporting both positive and negative expression [38,39]. Karlsson et al. [22] reported that hES-MP cells express CD10 when supplemented with a human serum. A review of the literature revealed several studies that evaluated cell surface markers expressed on hES-MP cells following expansion in FBS-supplemented media, none of which reported on the expression of CD10 [23,24,34,40,41]. CD10 has been identified as playing a role in maintaining the stem cell pool in various body compartments, especially the early progenitor population. This is achieved by its endopeptidase activity, which cleaves proteins in the cellular microenvironment and thus prevents them from promoting differentiation [42,43]. The expression of CD10 is associated with higher stemness of expressing cells when compared to nonexpressing cells [42]. Here, we demonstrated for the first time that hES-MP cells do not express CD10 after expansion in FBS-supplemented media, which possibly points to lower overall stemness of the hES-MP population following culture media supplementation with FBS. Further studies into factors causing CD10 expression, or lack thereof, based on the expansion media used are warranted.

The expression of both CD13 and CD105 was detected, regardless of whether FBS or hPL was used as a culture media supplement (Figure 2). The expression of CD13, however, was found to increase with time for the hPL–hES-MP cells. The CD105 expression was significantly lower in hPL–hES-MP cells than in the FBS–hES-MP cells. CD13 (also known as aminopeptidase N) has been described as a cellular marker for MSCs and other adult stem cells, even though its role in stem cell biology remains largely unknown [44]. Studies done on liver cancer stem cells and skeletal muscle satellite stem cells have demonstrated the role of CD13 in the survival of the stem cell pool [44,45]. Gabrilovac et al. [46] demonstrated that TGF-β increases the expression of CD13 on HL-60 cells in a concentration- and time-dependent manner. The time-dependent increase in CD13 expression seen in the hES-MP cells grown in the hPL-supplemented medium could be explained by the higher amount of TGF-β observed in hPL than in FBS (Figure S2). The significance of the CD13 expression in the hES-MP cells, however, remains unknown.

The stronger CD105 expression was associated with the hES-MP cells cultured in the medium supplemented with FBS. CD105 (also known as endoglin) is a co-receptor for ligands of the TGF-β family, such as TGF-β1 and BMP-2, and is important for activating the Smad signaling pathway [47,48]. The role of CD105 in cancers has been studied extensively and is commonly associated with migration and adhesion of cells, as well as being important for angiogenesis [47,49–51]. The role of CD105 in the cell biology of MSCs and hES-MP cells, however, has been studied only to a very limited extent and remains largely unknown [50,52].

It was found that the hES-MP cells successfully differentiated towards both osteoblasts and chondrocytes after expansion in culture media supplemented with hPL or with FBS. The ALP activity increased with time, indicating successful differentiation towards osteoblasts (Figure 3A). Tissue mineralization is normally seen at advanced stages of differentiation. The in vitro mineralization of MSCs is usually detected after 21 days of growth in the presence of osteogenic stimulants [53]. Our results for hES-MP cells are therefore consistent with previous findings for MSCs; in this study, hPL–hES-MP cells mineralized later (day 28) than FBS–hES-MP cells (day 21) (Figure 4). This may be related to the fact that hPL contains significantly higher levels of TGF-β than FBS. TGF-β has been described as an inhibitor of osteogenesis, promoting the development of pre-osteoblasts but inhibiting the development of mature osteoblasts and mineralization [54]. During differentiation, all hES-MP cell cultures were treated with the same osteogenic induction media, regardless of whether they were expanded in hPL or FBS, and thus high levels of TGF-β were not present at the time of differentiation. However, pre-treatment with high levels of TGF-β prior to differentiation might encourage the cells to mineralize later than when pre-treated with low levels of TGF-β.

The significantly higher expression of the osteogenic markers *RUNX2* and *SPP1* was observed in osteoblasts derived from hPL than in FBS-supplemented hES-MP cell cultures at all time points evaluated (Figure 5). The upregulation of osteogenic markers and the increased ALP activity during differentiation are both suggestive of successful osteoblast formation and indicate that growing hES-MP cells in hPL-enriched culture media allows them to preserve their osteogenic potential. How delayed mineralization will affect the quality of differentiation remains to be determined.

Chondrogenic differentiation was evaluated in terms of the accumulation of GAG and the expression of the chondrocytic marker *SOX9*. hES-MP cells cultured in the hPL-supplemented medium prior to differentiation initially expressed SOX9 at significantly higher levels than cells cultured in the FBS-supplemented medium (at 7 and 14 days), but this was reversed at later time points (Figure 5C). In all cultures, the *SOX9* expression peaked at day 28 and subsequently declined, similar to what has been reported by others [55]. However, at day 28, chondrocytes derived from hES-MP cells grown in the FBS-supplemented medium prior to differentiation had four-fold expression of *SOX9* compared to those derived from cells from the hPL-supplemented medium. GAG concentrations were initially observed to increase, but they subsequently decreased over a period of five weeks. No differences were observed, based on which culture supplement the cells were treated with prior to differentiation (Figure 3B). These findings are consistent with previous reports for MSC chondrogenic differentiation [25].

It should be noted that there was a slight difference in processing the hPL and FBS media supplements in this study: hPL was subjected to centrifugation and filtration steps, while FBS was not. This difference could contribute to the differences observed between cells derived in FBS- and hPL-supplemented culture media.

In the current study, the presence of extracellular vesicles (EVs) in hPL was not considered. The presence of EVs in hPL, such as fibrinogen, can result in gel formation in culture media as well as the precipitation of fibrin during storage and cell culture, and it may influence cell migration, proliferation, and differentiation [56]; therefore, EVs in the culture media supplements used in this study may influence the results to some degree. Therefore, in future work, the removal of EVs from hPL should be carried out (e.g., using the protocol developed by Pachler et al. [56]) and confirmed (e.g., via the monitoring of the platelet marker CD41).

5. Conclusions

hPLs are suitable for enriching culture media for the growth of hES-MP cells and allow cells to maintain good growth, osteogenic and chondrogenic differentiation, and normal expression of cell surface markers. Since cells are derived under animal-free conditions, hPL can be used instead of FBS as a growth medium supplement to ensure the continuity of an animal-free environment during cell culture.

Supplementary Materials: The following are available online at http://www.mdpi.com/2306-5354/7/3/75/s1, Figure S1: Human serum albumen concentrations for different hPL batches, Figure S2: Concentrations of key growth factors in hPL and FBS treatments, Figure S3: Morphology and trilineage differentiation of hES-MP cells, Figure S4: Surface marker expression on hES-MP cells following culture in hPL, Figure S5: hES-MP cartilage pellets after 28 days of differentiation.

Author Contributions: Conceptualization, O.E.S. and S.M.J.-B.; methodology, S.M.J.-B.; validation, S.M.J.-B. and O.E.S.; formal analysis, S.M.J.-B.; investigation, S.M.J.-B. and K.G.; resources, S.M.J.-B. and O.E.S.; data curation, O.E.S. and S.M.J.-B.; writing of the original draft preparation, S.M.J.-B.; writing of review and editing, O.E.S.; visualization, S.M.J.-B.; supervision, O.E.S.; project administration, O.E.S.; funding acquisition, O.E.S. All authors have read and agreed to the published version of the manuscript.

Funding: This research was partially supported by the IRF Fund (grant number 152388-052).

Acknowledgments: The authors want to thank the Landspitali University Hospital Research Fund for partially funding the project. We would also like to thank Ragna Landro and Bjorn Hardarsson for technical support and Kristine Wichuk and Steinunn Thorlacius for editing and reviewing the manuscript. We are grateful to the Blood Bank, Reykjavik, Iceland and Takara Bio Europe AB (previously Cellartis AB), Gothenburg, Sweden for providing the platelets and hES-MP cells, respectively, used in the study.

Conflicts of Interest: O.E.S. is an employee of the Icelandic Blood Bank at Landspitali University Hospital, the University of Reykjavik, and the University of Iceland. S.M.J.-B. and K.G. are graduated students at the University of Iceland. O.E.S. and S.M.J.-B. are shareholders in Platome Biotechnology.

References

1. Mountford, J.C. Human embryonic stem cells: Origins, characteristics and potential for regenerative therapy. *Transfus. Med.* **2008**, *18*, 1–12. [CrossRef] [PubMed]
2. Hanna, J.H.; Saha, K.; Jaenisch, R. Pluripotency and cellular reprogramming: Facts, hypotheses, unresolved issues. *Cell* **2010**, *143*, 508–525. [CrossRef] [PubMed]
3. de Peppo, G.M.; Marolt, D. State of the art in stem cell research: Human embryonic stem cells, induced pluripotent stem cells, and transdifferentiation. *J. Blood Transfus.* **2012**, *2012*, 317632. [CrossRef] [PubMed]
4. Kolios, G.; Moodley, Y. Introduction to stem cells and regenerative medicine. *Respiration* **2013**, *85*, 3–10. [CrossRef] [PubMed]
5. Fernández Vallone, V.B.; Romaniuk, M.A.; Choi, H.; Labovsky, V.; Otaegui, J.; Chasseing, N.A. Mesenchymal stem cells and their use in therapy: What has been achieved? *Differentiation* **2013**, *85*, 1–10. [CrossRef]
6. Richardson, S.M.; Kalamegam, G.; Pushparaj, P.N.; Matta, C.; Memic, A.; Khademhosseini, A.; Mobasheri, R.; Poletti, F.L.; Hoyland, J.A.; Mobasheri, A. Mesenchymal stem cells in regenerative medicine: Focus on articular cartilage and intervertebral disc regeneration. *Methods* **2016**, *99*, 69–80. [CrossRef]
7. Marquez-Curtis, L.A.; Janowska-Wieczorek, A.; McGann, L.E.; Elliott, J.A.W. Mesenchymal stromal cells derived from various tissues: Biological, clinical and cryopreservation aspects. *Cryobiology* **2015**, *71*, 181–197. [CrossRef]
8. Augello, A.; Kurth, T.B.; de Bari, C. Mesenchymal stem cells: A perspective from in vitro cultures to in vivo migration and niches. *Eur. Cells Mater.* **2010**, *20*, 121–133. [CrossRef]
9. Raggi, C.; Berardi, A.C. Mesenchymal stem cells, aging and regenerative medicine. *Muscles. Ligaments Tendons J.* **2012**, *2*, 239–242.
10. Zhang, J.; An, Y.; Gao, L.N.; Zhang, Y.J.; Jin, Y.; Chen, F.M. The effect of aging on the pluripotential capacity and regenerative potential of human periodontal ligament stem cells. *Biomaterials* **2012**, *33*, 6974–6986. [CrossRef]
11. Ikebe, C.; Suzuki, K. Mesenchymal stem cells for regenerative therapy: Optimization of cell preparation protocols. *Biomed Res. Int.* **2014**, *2014*, 951512. [CrossRef] [PubMed]

12. Lian, Q.; Lye, E.; Suan Yeo, K.; Khia Way Tan, E.; Salto-Tellez, M.; Liu, T.M.; Palanisamy, N.; El Oakley, R.M.; Lee, E.H.; Lim, B.; et al. Derivation of clinically compliant MSCs from CD105+, CD24− differentiated human ESCs. *Stem Cells* **2007**, *25*, 425–436. [CrossRef]
13. Trivedi, P.; Hematti, P. Derivation and immunological characterization of mesenchymal stromal cells from human embryonic stem cells. *Exp. Hematol.* **2008**, *36*, 350–359. [CrossRef] [PubMed]
14. Kopher, R.A.; Penchev, V.R.; Islam, M.S.; Hill, K.L.; Khosla, S.; Kaufman, D.S. Human embryonic stem cell-derived CD34+ cells function as MSC progenitor cells. *Bone* **2010**, *47*, 718–728. [CrossRef] [PubMed]
15. Sánchez, L.; Gutierrez-Aranda, I.; Ligero, G.; Rubio, R.; Muñoz-López, M.; García-Pérez, J.L.; Ramos, V.; Real, P.J.; Bueno, C.; Rodríguez, R.; et al. Enrichment of human ESC-derived multipotent mesenchymal stem cells with immunosuppressive and anti-inflammatory properties capable to protect against experimental inflammatory bowel disease. *Stem Cells* **2011**, *29*, 251–262. [CrossRef] [PubMed]
16. Gruenloh, W.; Kambal, A.; Sondergaard, C.; McGee, J.; Nacey, C.; Kalomoiris, S.; Pepper, K.; Olson, S.; Fierro, F.; Nolta, J.A. Characterization and in vivo testing of mesenchymal stem cells derived from human embryonic stem cells. *Tissue Eng.—Part A* **2011**, *17*, 1517–1525. [CrossRef] [PubMed]
17. Varga, N.; Veréb, Z.; Rajnavölgyi, É.; Német, K.; Uher, F.; Sarkadi, B.; Apáti, Á. Mesenchymal stem cell like (MSCl) cells generated from human embryonic stem cells support pluripotent cell growth. *Biochem. Biophys. Res. Commun.* **2011**, *414*, 474–480. [CrossRef]
18. Brown, P.T.; Squire, M.W.; Li, W.J. Characterization and evaluation of mesenchymal stem cells derived from human embryonic stem cells and bone marrow. *Cell Tissue Res.* **2014**, *358*, 149–164. [CrossRef]
19. Amit, M.; Shariki, C.; Margulets, V.; Itskovitz-Eldor, J. Feeder layer- and serum-free culture of human embryonic stem cells. *Biol. Reprod.* **2004**, *70*, 837–845. [CrossRef]
20. Hoffman, L.M.; Carpenter, M.K. Characterization and culture of human embryonic stem cells. *Nat. Biotechnol.* **2005**, *23*, 699–708. [CrossRef]
21. Sánchez, L.; Gutierrez-Aranda, I.; Ligero, G.; Martín, M.; Ayllón, V.; Real, P.J.; Ramos-Mejía, V.; Bueno, C.; Menendez, P. Maintenance of human embryonic stem cells in media conditioned by human mesenchymal stem cells obviates the requirement of exogenous basic fibroblast growth factor supplementation. *Tissue Eng. Part C Methods* **2012**, *18*, 387–396. [CrossRef] [PubMed]
22. Karlsson, C.; Emanuelsson, K.; Wessberg, F.; Kajic, K.; Axell, M.Z.; Eriksson, P.S.; Lindahl, A.; Hyllner, J.; Strehl, R. Human embryonic stem cell-derived mesenchymal progenitors—Potential in regenerative medicine. *Stem Cell Res.* **2009**, *3*, 39–50. [CrossRef] [PubMed]
23. de Peppo, G.M.; Sladkova, M.; Sjövall, P.; Palmquist, A.; Oudina, K.; Hyllner, J.; Thomsen, P.; Petite, H.; Karlsson, C. Human embryonic stem cell-derived mesodermal progenitors display substantially increased tissue formation compared to human mesenchymal stem cells under dynamic culture conditions in a packed bed/column bioreactor. *Tissue Eng. Part A* **2013**, *19*, 175–187. [CrossRef] [PubMed]
24. Li, O.; Tormin, A.; Sundberg, B.; Hyllner, J.; Le Blanc, K.; Scheding, S. Human embryonic stem cell-stroma cells (hES-MSCs) engraft in vivo and support hematopoiesis without suppressing immune function: Implications for off-the shelf ES-MSC therapies. *PLoS ONE* **2013**, *8*, e55319. [CrossRef]
25. Jonsdottir-Buch, S.M.; Lieder, R.; Sigurjonsson, O.E. Platelet lysates produced from expired platelet concentrates support growth and osteogenic differentiation of mesenchymal stem cells. *PLoS ONE* **2013**, *8*, e68984. [CrossRef]
26. Jonsdottir-Buch, S.M.; Sigurgrimsdottir, H.; Lieder, R.; Sigurjonsson, O.E. Expired and pathogen-inactivated platelet concentrates support differentiation and immunomodulation of mesenchymal stromal cells in culture. *Cell Transplant.* **2015**, *24*, 1545–1554. [CrossRef]
27. Schallmoser, K.; Bartmann, C.; Rohde, E.; Reinisch, A.; Kashofer, K.; Stadelmeyer, E.; Drexler, C.; Lanzer, G.; Linkesch, W.; Strunk, D. Human platelet lysate can replace fetal bovine serum for clinical-scale expansion of functional mesenchymal stromal cells. *Transfusion* **2007**, *47*, 1436–1446. [CrossRef]
28. Bieback, K.; Hecker, A.; Kocaömer, A.; Lannert, H.; Schallmoser, K.; Strunk, D.; Klüter, H. Human alternatives to fetal bovine serum for the expansion of mesenchymal stromal cells from bone marrow. *Stem Cells* **2009**, *27*, 2331–2341. [CrossRef]
29. Williamson, L.M.; Devine, D.V. Challenges in the management of the blood supply. *Lancet* **2013**, *381*, 1866–1875. [CrossRef]

30. Fernandez-Rebollo, E.; Mentrup, B.; Ebert, R.; Franzen, J.; Abagnale, G.; Sieben, T.; Ostrowska, A.; Hoffmann, P.; Roux, P.-F.; Rath, B.; et al. Human platelet lysate versus fetal calf serum: These supplements do not select for different mesenchymal stromal cells. *Sci. Rep.* **2017**, *7*, 5132. [CrossRef]
31. Kandoi, S.; Kumar, P.L.; Patra, B.; Vidyasekar, P.; Sivanesan, D.; Shanker Verma, R. Evaluation of platelet lysate as a substitute for FBS in explant and enzymatic isolation methods of human umbilical cord MSCs. *Sci. Rep.* **2018**, *8*, 12439. [CrossRef] [PubMed]
32. Karadjian, M.; Senger, A.S.; Essers, C.; Wilkesmann, S.; Heller, R.; Fellenberg, J.; Simon, R.; Westhauser, F. Human platelet lysate can replace fetal calf serum as a protein source to promote expansion and osteogenic differentiation of human bone-marrow-derived mesenchymal stromal cells. *Cells* **2020**, *9*, 918. [CrossRef]
33. Pfaffl, M.W.; Horgan, G.W.; Dempfle, L. Relative expression software tool (REST©) for group-wise comparison and statistical analysis of relative expression results in real-time PCR. *Nucleic Acids Res.* **2002**, *30*, e36. [CrossRef]
34. de Peppo, G.M.; Svensson, S.; Lennerås, M.; Synnergren, J.; Stenberg, J.; Strehl, R.; Hyllner, J.; Thomsen, P.; Karlsson, C. Human embryonic mesodermal progenitors highly resemble human mesenchymal stem cells and display high potential for tissue engineering applications. *Tissue Eng. Part A* **2010**, *16*, 2161–2182. [CrossRef] [PubMed]
35. Burnouf, T.; Strunk, D.; Koh, M.B.C.; Schallmoser, K. Human platelet lysate: Replacing fetal bovine serum as a gold standard for human cell propagation? *Biomaterials* **2016**, *76*, 371–387. [CrossRef] [PubMed]
36. Dominici, M.; Le Blanc, K.; Mueller, I.; Slaper-Cortenbach, I.; Marini, F.C.; Krause, D.S.; Deans, R.J.; Keating, A.; Prockop, D.J.; Horwitz, E.M. Minimal criteria for defining multipotent mesenchymal stromal cells. The International Society for Cellular Therapy position statement. *Cytotherapy* **2006**, *8*, 315–317. [CrossRef]
37. Niehage, C.; Steenblock, C.; Pursche, T.; Bornhäuser, M.; Corbeil, D.; Hoflack, B. The cell surface proteome of human mesenchymal stromal cells. *PLoS ONE* **2011**, *6*, e20399. [CrossRef]
38. Javazon, E.H.; Beggs, K.J.; Flake, A.W. Mesenchymal stem cells: Paradoxes of passaging. *Exp. Hematol.* **2004**, *32*, 414–425. [CrossRef] [PubMed]
39. Mafi, P.; Hindocha, S.; Mafi, R.; Griffin, M.; Khan, W.S. Adult mesenchymal stem cells and cell surface characterization—A systematic review of the literature. *Open Orthop. J.* **2011**, *5*, 253–260. [CrossRef]
40. Lai, R.C.; Arslan, F.; Tan, S.S.; Tan, B.; Choo, A.; Lee, M.M.; Chen, T.S.; Teh, B.J.; Eng, J.K.L.; Sidik, H.; et al. Derivation and characterization of human fetal MSCs: An alternative cell source for large-scale production of cardioprotective microparticles. *J. Mol. Cell. Cardiol.* **2010**, *48*, 1215–1224. [CrossRef]
41. Chen, W.; Zhou, H.; Weir, M.D.; Tang, M.; Bao, C.; Xu, H.H.K. Human embryonic stem cell-derived mesenchymal stem cell seeding on calcium phosphate cement-chitosan-RGD scaffold for bone repair. *Tissue Eng. Part A* **2013**, *19*, 915–927. [CrossRef] [PubMed]
42. Bachelard-Cascales, E.; Chapellier, M.; Delay, E.; Pochon, G.; Voeltzel, T.; Puisieux, A.; Caron de Fromentel, C.; Maguer-Satta, V. The CD10 enzyme is a key player to identify and regulate human mammary stem cells. *Stem Cells* **2010**, *28*, 1081–1088. [CrossRef] [PubMed]
43. Maguer-Satta, V.; Besançon, R.; Bachelard-Cascales, E. Concise review: Neutral endopeptidase (CD10): A multifaceted environment actor in stem cells, physiological mechanisms, and cancer. *Stem Cells* **2011**, *29*, 389–396. [CrossRef]
44. Rahman, M.M.; Ghosh, M.; Subramani, J.; Fong, G.-H.; Carlson, M.E.; Shapiro, L.H. CD13 regulates anchorage and differentiation of the skeletal muscle satellite stem cell population in ischemic injury. *Stem Cells* **2014**, *32*, 1564–1577. [CrossRef] [PubMed]
45. Kim, H.M.; Haraguchi, N.; Ishii, H.; Ohkuma, M.; Okano, M.; Mimori, K.; Eguchi, H.; Yamamoto, H.; Nagano, H.; Sekimoto, M.; et al. Increased CD13 expression reduces reactive oxygen species, promoting survival of liver cancer stem cells via an epithelial-mesenchymal transition-like phenomenon. *Ann. Surg. Oncol.* **2012**, *19*, 539–548. [CrossRef] [PubMed]
46. Gabrilovac, J.; Breljak, D.; Čupić, B. Regulation of aminopeptidase N (EC 3.4.11.2; APN; CD13) on the HL-60 cell line by TGF-β1. *Int. Immunopharmacol.* **2008**, *8*, 613–623. [CrossRef]
47. Pérez-Gómez, E.; Del Castillo, G.; Santibáñez, J.F.; López-Novoa, J.M.; Bernabéu, C.; Quintanilla, M. The role of the TGF-β coreceptor endoglin in cancer. *Sci. World J.* **2010**, *10*, 2367–2384. [CrossRef]
48. Nassiri, F.; Cusimano, M.D.; Scheithauer, B.W.; Rotondo, F.; Fazio, A.; Yousef, G.M.; Syro, L.V.; Kovacs, K.; Lloyd, R.V. Endoglin (CD105): A review of its role in angiogenesis and tumor diagnosis, progression and therapy. *Anticancer Res.* **2011**, *31*, 2283–2290.

49. Fonsatti, E.; Maio, M. Highlights on endoglin (CD105): From basic findings towards clinical applications in human cancer. *J. Transl. Med.* **2004**, *2*, 18. [CrossRef]
50. Dallas, N.A.; Samuel, S.; Xia, L.; Fan, F.; Gray, M.J.; Lim, S.J.; Ellis, L.M. Endoglin (CD105): A marker of tumor vasculature and potential target for therapy. *Clin. Cancer Res.* **2008**, *14*, 1931–1937. [CrossRef]
51. Todorović-Raković, N.; Milovanović, J.; Nikolić-Vukosavljević, D. TGF-β and its coreceptors in cancerogenesis: An overview. *Biomark. Med.* **2011**, *5*, 855–863. [CrossRef] [PubMed]
52. Fonsatti, E.; Del Vecchio, L.; Altomonte, M.; Sigalotti, L.; Nicotra, M.R.; Coral, S.; Natali, P.G.; Maio, M. Endoglin: An accessory component of the TGF-β-binding receptor-complex with diagnostic, prognostic, and bioimmunotherapeutic potential in human malignancies. *J. Cell. Physiol.* **2001**, *188*, 1–7. [CrossRef] [PubMed]
53. Chevallier, N.; Anagnostou, F.; Zilber, S.; Bodivit, G.; Maurin, S.; Barrault, A.; Bierling, P.; Hernigou, P.; Layrolle, P.; Rouard, H. Osteoblastic differentiation of human mesenchymal stem cells with platelet lysate. *Biomaterials* **2010**, *31*, 270–278. [CrossRef] [PubMed]
54. Matsumoto, T.; Abe, M. TGF-β-related mechanisms of bone destruction in multiple myeloma. *Bone* **2011**, *48*, 129–134. [CrossRef] [PubMed]
55. Yamashita, A.; Nishikawa, S.; Rancourt, D.E. Identification of five developmental processes during chondrogenic differentiation of embryonic stem cells. *PLoS ONE* **2010**, *5*, e10998. [CrossRef]
56. Pachler, K.; Lener, T.; Streif, D.; Dunai, Z.A.; Desgeorges, A.; Feichtner, M.; Öller, M.; Schallmoser, K.; Rohde, E.; Gimona, M. A Good Manufacturing Practice–grade standard protocol for exclusively human mesenchymal stromal cell–derived extracellular vesicles. *Cytotherapy* **2017**, *19*, 458–472. [CrossRef]

 *bioengineering*

Article

Bioreactor Parameters for Microcarrier-Based Human MSC Expansion under Xeno-Free Conditions in a Vertical-Wheel System

Josephine Lembong, Robert Kirian, Joseph D. Takacs, Timothy R. Olsen, Lye Theng Lock, Jon A. Rowley and Tabassum Ahsan *

RoosterBio, Inc., 5295 Westview Drive, Suite 275, Frederick, MD 21703, USA; josephine@roosterbio.com (J.L.); robert@roosterbio.com (R.K.); joseph@roosterbio.com (J.D.T.); tim@roosterbio.com (T.R.O.); lyetheng@gmail.com (L.T.L.); jon@roosterbio.com (J.A.R.)

* Correspondence: taby@roosterbio.com

Received: 2 June 2020; Accepted: 5 July 2020; Published: 8 July 2020

Abstract: Human mesenchymal stem/stromal cells (hMSCs) have been investigated and proven to be a well-tolerated, safe therapy for a variety of indications, as shown by over 900 registered hMSC-based clinical trials. To meet the commercial demand for clinical manufacturing of hMSCs, production requires a scale that can achieve a lot size of ~100B cells, which requires innovative manufacturing technologies such as 3D bioreactors. A robust suspension bioreactor process that can be scaled-up to the relevant scale is therefore crucial. In this study, we developed a fed-batch, microcarrier-based bioreactor process, which enhances media productivity and drives a cost-effective and less labor-intensive hMSC expansion process. We determined parameter settings for various stages of the culture: inoculation, bioreactor culture, and harvest. Addition of a bioreactor feed, using a fed-batch approach, was necessary to replenish the mitogenic factors that were depleted from the media within the first 3 days of culture. Our study resulted in an optimized hMSC culture protocol that consistently achieved hMSC densities between 2×10^5–6×10^5 cells/mL within 5 days with no media exchange, maintaining the final cell population doubling level (PDL) at 16–20. Using multiple hMSC donors, we showed that this process was robust and yielded hMSCs that maintained expansion, phenotypic characteristic, and functional properties. The developed process in a vertical-wheel suspension bioreactor can be scaled to the levels needed to meet commercial demand of hMSCs.

Keywords: bioreactor; hMSCs; microcarrier; bioprocess

1. Introduction

Human mesenchymal stem/stromal cells (hMSCs) have been investigated and proven to be a well-tolerated, safe therapy for a variety of diseases [1], as indicated by over 900 registered hMSC-based clinical trials in ClinicalTrials.gov as of 2019. Previous analysis of cell dose estimates from the ClinicalTrials.gov website shows that most indications requiring hMSC treatment need a target production lot size of ~100B cells [1,2], with some exceptions including low dose indications such as ocular diseases [3]. Such production lot size requires a scalable manufacturing platform such as 3D bioreactors, where MSC expansion in suspension is typically performed by growing cells on adherent substrates such as microcarriers [4–6]. A robust suspension bioreactor process that can be scaled-up is therefore crucial to meet this demand for clinical manufacturing of hMSCs, and this starts with process development and optimization at a scaled-down bioreactor model.

To accommodate hMSC expansion in suspension bioreactors, it is critical to choose a bioreactor platform that allows cell attachment to the microcarriers, homogeneously suspends the cells and microcarriers, as well as distributes gas and nutrients while imposing low shear, as hMSCs are known to be sensitive to shear stress [7–9]. In addition to providing large cell numbers, an hMSC manufacturing platform should facilitate maintenance of the hMSC undifferentiated cell state and phenotype to retain their therapeutic potency. Among many bioreactor platforms, the stirred-tank bioreactor has been extensively evaluated for hMSC expansion [10–13]. More recently, vertical-wheel bioreactors have been developed to provide a uniform, low-shear fluid mixing environment for efficient particle suspension. The agitation mechanics provided by the vertical impeller and the shape of the bioreactor vessel allows for a better homogenization of the microcarriers with low power input compared to commonly used cylindrical bioreactors with horizontal impellers [14]. The low power, and consequently low hydrodynamic shear, makes this bioreactor specifically a great candidate for culture of shear-sensitive, anchorage dependent cells such as hMSCs [15,16]. In addition, this low shear environment remains constant across the full range of vessel sizes from 0.5 to 500 L [17]. For hMSC culture, it has been reported that the vertical-wheel bioreactor facilitated higher cell-microcarrier attachment compared to the stirred-tank bioreactor and the generated hMSCs showed a higher percentage of proliferative cells, lower percentage of apoptotic cells, and reduced levels of HLA-DR positive cells compared to cells grown on stirred-tank bioreactors [18]. The availability of this bioreactor platform in multiple scales, currently ranging from 0.1 L to 80 L, is also key as hMSC manufacturing processes need to be developed with future scalability in mind. Thus, a low shear, vertical-wheel bioreactor was chosen as the platform to develop our process.

In clinical hMSC manufacturing, there has been movement away from serum-containing processes due to higher risk of graft rejection, immunoreactions, and virus contamination associated with animal-derived products [19–21]. To ensure product safety, xeno-free (XF) supplements such as human serum and human platelet lysate [19,22–24] have therefore become increasingly incorporated in hMSC clinical manufacturing processes [21]. While XF MSC-microcarrier cultures have been developed, it typically involves full or partial media exchanges to manage nutrient supply and waste build-up [25–29]. Borrowing from best practices in the monoclonal antibody field, fed batch processes can lead to high media productivity [30], as defined in the Glossary for Cell & Gene Therapy and Regenerative Medicine as millions of cells/liter of media consumed [31]. Optimizing this media productivity is critical as media is the major cost driver in cell manufacturing [30], which also carries the hidden costs of labor associated with media exchange.

Another paradigm for consideration is perfusion, in which media is continuously circulated through the culture, therefore simultaneously supplying nutrients and removing waste from the culture. While perfusion offers a continuous process and can achieve high cell densities, it utilizes large quantities of media and consequently induces high media costs and logistics/supply chain issues [32]. Thus, perfusion has not been widely adopted at larger scales. In addition, most hMSC culture process development and pilot scale manufacturing utilize a fed-batch paradigm [33], making it a bigger regulatory hurdle to switch to a perfusion process and demonstrate comparability [32]. For this reason, a fed-batch process was developed in this study with future scalability in mind.

We established this fed-batch process in hMSC expansion in the vertical-wheel bioreactors, where a fed-batch process showed a higher cell yield and therefore enhanced media productivity compared to a batch process (Figure 1A,B). Addition of the bioreactor feed was necessary to replenish the growth factors that were depleted from the media within the first 3 days of culture (Figure 1A). Formation of cell-bead aggregates during the expansion process was observed (Figure 1C), with aggregate size increasing over time.

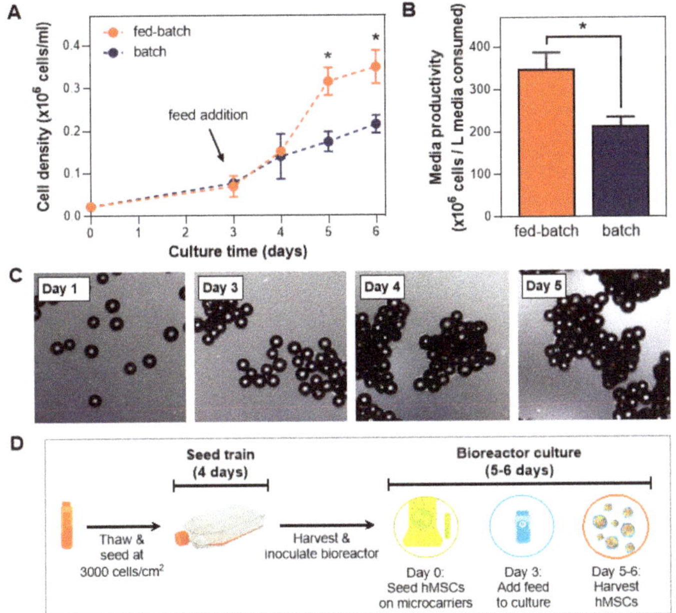

Figure 1. Fed-batch bioreactor culture paradigm. (**A**) Comparison study of a xeno-free (XF) bioreactor process utilizing batch and fed-batch processes shows a distinct advantage of the fed-batch process on final cell yield (n = 3). (**B**) Media productivity is enhanced using a fed-batch process. (**C**) Images indicating typical cell growth on microcarriers during a 5-day culture in a vertical-wheel bioreactor (* $p \leq 0.05$). (**D**) Schematic of the bioreactor culture workflow using a fed-batch process.

With the fed-batch paradigm in place, further optimization of this MSC bioreactor culture process is central to maximizing yields and recovering healthy, functional cells at harvest. To achieve that, optimization studies need to be performed in the small-scale bioreactors prior to scaling up to a full production scale. Various stages of a XF bioreactor culture need to be studied: cell and microcarrier inoculation, seeding strategies, culture, feed, and harvest protocol (Figure 1D). Each of these steps is associated with numerous parameters that need to be evaluated, which is the objective of our study. In this paper, we developed and optimized a xeno-free, fed-batch, microcarrier-based culture process in a scalable low-shear bioreactor system, resulting in consistent hMSC yield and maintenance of cell health and characteristics following culture.

2. Materials and Methods

2.1. hMSC Planar Culture for Seed Train

hBM-MSC Working Cell Banks (WCB) from five adult donors (RoosterBio, Frederick, MD, USA) were used in this study. These cells had been isolated and previously expanded under XF conditions (PDL: 8-10, Passage number: 2). Cells were thawed and seeded at a density of 3000 cells/cm^2 in Corning® CellBIND® culture flasks (Corning Incorporated, Corning, NY, USA), in the accompanying High Performance XF Media Kit (RoosterBio). Cells were cultured at 37 °C with 5% CO_2. After 4 days, cells were dissociated from the flasks with TrypLE™ Select Enzyme (Thermo Fisher Scientific, Waltham, MA, USA), quenched with medium, then counted using the NucleoCounter® NC-200™ (Chemometec, Allerod, Denmark). Cell suspension was then spun down at 300× g for 5 min and resuspended in growth medium at a density of 1×10^6 cells/mL for cell inoculation in the following steps.

2.2. hMSC Bioreactor Culture

Following the seed train, hMSCs were cultured on Corning® Low Concentration Synthemax™ II microcarriers (Corning) inside PBS-0.1 single-use Vertical-Wheel™ bioreactors (PBS Biotech, Camarillo, CA, USA). To seed the cells onto the microcarriers, a 1–6.3 mL cell suspension was inoculated into the bioreactor containing 30 mL of medium and 0.75–2.5 g of microcarriers, resulting in cell seeding density of 11,000–70,000 cells/mL, and areal density of 2222–9333 cells/cm^2 of microcarrier surface. To facilitate attachment, the cells were incubated for 20 min at static condition, then gently shaken to re-distribute the cells and the microcarriers, and finally incubated for 20 more minutes to further cell attachment to microcarriers. Our data showed that an average of 84% cell attachment could be achieved using this seeding method (Figure S1).

At the end of the cell attachment duration, fresh culture medium was added to a total volume of 90 mL. The PBS-0.1 vessel was placed on its controller base in a 5% CO_2 incubator and agitation was initiated at 25 rpm. On Day 3 of the bioreactor culture, an XF bioreactor feed, RoosterReplenish™-MSC-XF (RoosterBio), was added at 2% of the working volume and agitation speed was increased to 30 rpm.

2.3. Bioreactor Sampling

Daily sampling of the bioreactor was performed to quantify cell numbers as well as the metabolite and waste profile to assess the kinetics of cell growth and media consumption. A 3 mL sample was taken daily while the cells/microcarriers were in suspension at 30 rpm agitation. The cell/microcarrier sample were left to sediment, the supernatant was isolated for medium analysis, and an equal volume of TrypLE was added to dissociate the cells from the microcarriers. Following dissociation, the number of cells was counted. The medium supernatant was analyzed using the BioProfile FLEX2 Analyzer (Nova Biomedical, Waltham, MA, USA) to monitor the metabolite and waste concentration.

Visualization of cell attachment and cell/microcarrier aggregation was performed by obtaining a 1 mL sample from the bioreactor and visualizing it under light microscopy (EVOS M5000 Cell Imaging System; ThermoFisher Scientific).

2.4. In-Vessel Harvest

Following hMSC expansion in bioreactors for 5–6 days, in-vessel harvest was performed. While bioreactor sampling provides the kinetics during culture, the most accurate quantification of total cell yield after the culture duration is obtained through complete in-vessel harvest, minimizing any potential errors from non-representative sampling when large cell-microcarrier aggregates are present in the culture. After agitation is stopped and the cells/microcarriers settle to the bottom of the bioreactor, the medium was removed and the remaining cells/microcarrier solution were washed with PBS. TrypLE was added to the bioreactor at half of the working volume, and agitation was initiated (20–100 rpm for 20–40 min) in a 37 °C incubator. Cells were then separated from the microcarriers using a 100-µm cell strainer, quenched with an equal volume of medium, and then counted to obtain the final harvest yield. Values are presented as concentrations (cells/mL) to emphasize the scalable process that was being developed.

2.5. Analysis of Cell Health, Quality Attributes and Functionality

Evaluation of hMSC properties was conducted on the cells harvested from the bioreactor and compared to the properties of cells that were expanded on 2D planar flasks, which served as a control. Cell populations were evaluated for expression of MSC surface markers and differentiation potential (to the adipo-, osteo-, and chondrogenic lineages). Functional properties of MSCs were also evaluated, such as immunomodulatory function and angiogenic cytokine secretion.

2.5.1. Cell Health Analysis

To determine the health of the hMSCs harvested from the bioreactor, cell expansion was quantified by culturing the cells harvested from the bioreactor in CellBIND flasks for 5 days. Following a 5-day culture, cells were harvested and the fold expansion was quantified. Expansion capability of the cells harvested from the bioreactor is an important measure of cell quality/health and is an important validation of the bioreactor process following the final cell yield.

2.5.2. Tri-Lineage Differentiation

Qualitative evaluation of hMSC differentiation down adipo-, osteo-, and chondrogenic lineages were performed on the cells harvested from the bioreactor and compared to those from the 2D flask controls. For adipogenic differentiation, harvested cells were cultured in STEMPRO® Adipogenesis Differentiation Kit media (Gibco) for 7–14 days, then fixed and stained with Oil Red O Solution (Sigma-Aldrich). For osteogenic differentiation, cells were cultured in STEMPRO® Osteogenesis Differentiation Kit media (Gibco) for 14 days, then fixed and stained with 2% Alizarin Red Stain (Lifeline Cell Technology, Walkersville, MD, USA). For chondrogenic differentiation, cell pellets were generated in ultra-low attachment plates and cultured in STEMPRO® Chondrogenesis Differentiation Kit media (Gibco) for 21 days, then fixed, sectioned, and stained with Alcian Blue. All differentiated samples were imaged following staining and compared with matched undifferentiated samples.

2.5.3. Flow Cytometry for Cell Surface Marker Expression

Cells dissociated from the microcarriers and harvested from the bioreactors were analyzed for the expression of surface markers (CD73(+), CD90(+), CD105(+), CD166(+), CD14(−), CD34(−), CD45(−)) using flow cytometry and compared to those from 2D flask controls.

2.5.4. Immunomodulatory Function

Induction of indoleamine 2,3-dioxygenase (IDO) activity by exposure of hMSCs to the pro-inflamatory cytokine interferon-gamma (IFN-γ) is central to the immunosuppressive function of hMSCs [34–36]. hBM-MSCs harvested from the bioreactors were cultured for 24 ± 4 h, then washed and stimulated with 10 ng/mL IFN-γ for 24 h. The cell supernatant was collected, and the kynurenine concentration was measured using a spectrophotometric assay and normalized to number of cells and days of incubation to obtain the amount of IDO secreted (expressed as pg kynurenine per cell per day). This kynurenine level was then compared to the levels from the 2D flask controls.

2.5.5. Angiogenic Cytokine Secretion

MSCs have been shown to secrete cytokines which help promote tissue angiogenesis [37–39], therefore we quantify the secretion of various angiogenic cytokine as a measure of cell functionality. hBM-MSCs harvested from the bioreactors were cultured for 24 ± 4 h, then washed and switched to medium supplemented with 2% FBS. After 24-h incubation, the supernatant was collected and assayed for FGF, HGF, IL-8, TIMP-1, TIMP-2 and VEGF concentration using a MultiPlex ELISA (Quansys Biosciences, Logan, UT, USA). Cytokine concentration was normalized to number of cells and days of incubation to obtain cytokine secretion rates, and compared to the concentrations from the 2D flask controls.

2.6. Statistical Analysis

We previously quantified bioreactor-to-bioreactor variability within an experiment, and observed the following coefficient of variation values: CV < 6% for final harvest density and CV < 2% for post-bioreactor cell expansion (Figure S2). Due to these small variabilities, a ≥ 10% difference among groups of different experimental parameters was deemed biologically meaningful. A sample size of

$n = 1$ bioreactor per group in these kinetic experiments therefore still allowed for the determination of the effects of the tested experimental parameters. This approach allowed for the breadth of experiments presented here.

3. Results and Discussion

3.1. Cell and Microcarrier Inoculation Concentration in a Fixed Media Volume

To first determine the maximum expansion capability in a fixed media volume in the bioreactor, we performed a fed-batch culture in bioreactors where a range of cell numbers and microcarriers amount were seeded, from 1.25 g and 2.1×10^6 cells (1×) to three times the microcarriers and cell number at 3.75 g and 6.3×10^6 cells (3×). At 1.25 g, the microcarriers provide 450 cm^2 surface area in 90 mL medium, which is the recommended surface area to media volume ratio by the manufacturer. The cell density for the 1× condition is 23,333 cells/mL (4667 cells/cm^2). The effect of media exchanges on the group with the highest cell number and microcarriers (at a concentration that far exceed the manufacturer's recommendations) was also investigated (denoted as 3×ΔΔ in Figure 2). Among the three fed-batch conditions, we observed that bioreactor seeding with 2.1×10^6 cells with 1.25 g microcarriers resulted in the highest cell yield (Figure 2A,B). In the groups where 4.2×10^6 cells or 6.3×10^6 cells were seeded, cell growth was limited by the media, as shown by the decrease in cell density on day 3 or day 4 (Figure 2A). With two complete media exchanges on day 3 and day 4 (instead of the bioreactor feed addition), growth of 6.3×10^6 cells can be supported (Figure 2A, dotted line), justifying the media limitation in our fed-batch system at this high cell-to-media ratio. Media exchanges, however, are costly [30], resulting in low media productivity (Figure 2C), and completely untenable at larger bioreactor scales. Thus, a fed-batch paradigm is critical due to its high media productivity (Figure 1B). We therefore selected 2.1×10^6 cells with 1.25 g microcarriers for seeding, which then remained consistent during evaluation of subsequent process parameters.

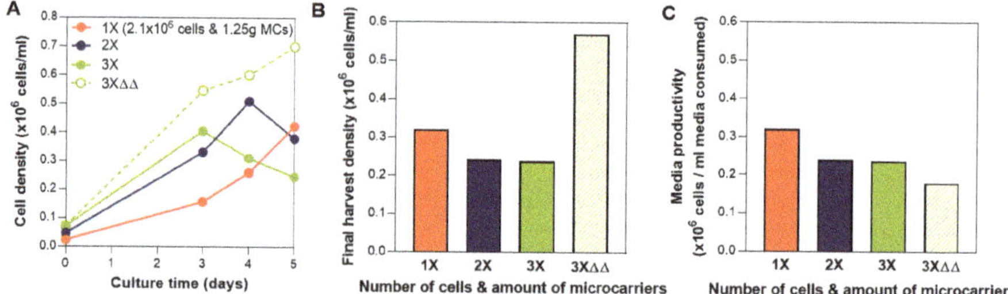

Figure 2. Determination of cell and microcarrier inoculation number. (**A**) Growth kinetics in bioreactors for various cell numbers and microcarrier during seeding, as well as media exchanges. Cell counts were obtained from sampling. (**B**) Final harvest density obtained from in-vessel harvest. At the highest tested cell-to-media ratio, cell growth can only be supported through media exchanges. (**C**) Media productivity comparison shows the disadvantage of media exchanges despite the high cell yield. Tested groups are 1×: 2.1×10^6 cells and 1.25 g; 2×: 4.2×10^6 cells and 2.5 g; 3×: 6.3×10^6 cells and 3.75 g; and 3×ΔΔ: 6.3×10^6 cells and 3.75 g with media exchanges on Days 3 and 4.

3.2. Bioreactor Seeding Strategies: Cell and Microcarrier Concentrations at Inoculation

The effects of cell seeding parameters, i.e., cell numbers and microcarrier surface area, were each investigated independently. First, various cell numbers were seeded into the bioreactor with 1.25 g microcarriers: 1×10^6 cells (2222 cells/cm^2 seeding density), 2.1×10^6 (4667 cells/cm^2 seeding density), and 4.2×10^6 cells (9333 cells/cm^2 seeding density). Seeding of 2.1×10^6 and 4.2×10^6 cells resulted in similar yield of 4×10^5 cells/mL, while bioreactor seeding with 1×10^6 cells resulted in the lowest yield (Figure 3(Ai,Aii)). However, a high cell number at inoculation (4.2×10^6 cells) reached a growth plateau after 3 days despite the addition of a bioreactor feed on Day 3 (Figure 3(Ai)). Note that it is best practice to harvest cells during the exponential growth phase while proliferation still occurs (not during the plateau) to help ensure the maintenance of the hMSC phenotype. Inoculation with 2.1×10^6 cells resulted in both the highest observed cell yield and maintained post-bioreactor cell health, as shown by the expansion of the cells harvested from the bioreactor, which is similar to that of the 2D control (Figure 3(Aiii)). Higher post-bioreactor expansion potential of the cells from the 1M group is likely due to the observed lower proliferation within the bioreactor, resulting in cells with a lower population double level (PDL). Medium analysis from all three bioreactor conditions revealed observable glucose consumption and lactate production over the culture duration (Figure 3(Aiv)), with the 4.2×10^6 group with the highest nutrient consumption and waste production. Taken together, these results indicated the use of 2.1×10^6 cells (23,333 cells/mL) as the cell inoculation density in subsequent studies, as after 5 days of culture the cell yield is maximized, and cell health is maintained.

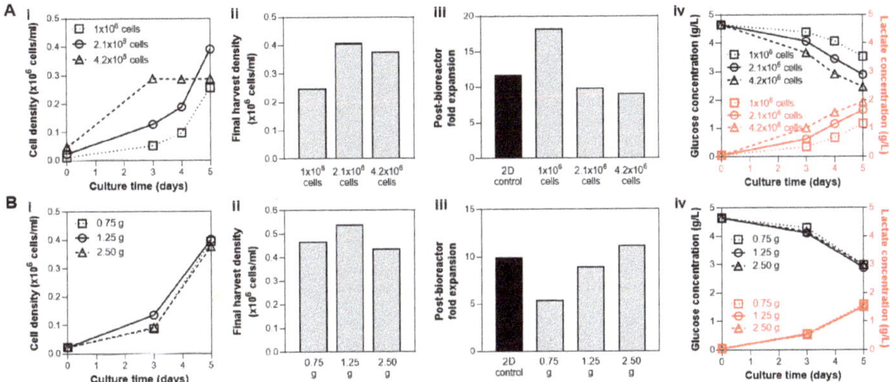

Figure 3. Optimization of bioreactor seeding parameters at fixed microcarrier amount (A; 1.25 g), or fixed cell number (B; 2.1×10^6 cells). (**A**) Effect of cell seeding numbers on bioreactor culture yield with 1.25 g microcarriers, and (**B**) effect of microcarrier density on bioreactor culture yield with 2.1×10^6 cells seeded. For both optimizations, we presented hMSC growth kinetics in bioreactors based on cell density from daily sampling (i), final density obtained from an in-vessel harvest (ii), expansion of cells harvested from the bioreactors compared to the 2D control (iii), and analysis of nutrient and waste in medium during culture (iv). Both seeding parameter studies support bioreactor seeding with 1.25 g of microcarriers and 2.1×10^6 cells (4667 cells/cm^2) to optimize for the cell yield and post-bioreactor expansion, while minimizing costs associated with materials.

With the cell to media ratio established at 2.1×10^6 cells and 90 mL, the effect of seeding density was then investigated. The amount of microcarriers were varied, providing a range of available surface area for cell growth: 0.75 g (270 cm^2), 1.25 g (450 cm^2), and 2.5 g (900 cm^2), which corresponds to seeding density of 7777 cells/cm^2, 4667 cells/cm^2, and 2333 cells/cm^2, respectively. The growth kinetics and final cell yield after a 5-day culture were similar among the three conditions (Figure 3(Bi,Bii)). Analysis of nutrient and waste concentration in the medium also showed comparable glucose consumption/lactate production profiles in all three bioreactor conditions (Figure 3(Biv)). However, the cells harvested from the bioreactor with 0.75 g microcarriers (7777 cells/cm^2) showed markedly reduced expansion compared to the other conditions with 1.25 g (4667 cells/cm^2) and 2.5 g microcarriers (2333 cells/cm^2), as well as the 2D control (Figure 3(Biii)). Thus, although bioreactor inoculation with 2.5 g microcarriers resulted in similar biological metrics (final harvest density, post-bioreactor cell expansion, and metabolite profile), 1.25 g was preferred due to the reduction in materials cost, which has notable impact when processes are scaled up to the pilot and production level (potentially hundreds to thousands of liters).

These results indicate the use of 2.1×10^6 cells and 1.25 g microcarriers for bioreactor inoculation. The formation of hMSC-microcarrier aggregates associated with these bioreactor seeding studies is shown in Figure S3, where aggregate size at Day 5 was seen to be independent of cell numbers or microcarrier amount at seeding. This qualitative assessment of the bioreactor culture looks largely similar for all of the experiments; therefore, typical pictures are not shown for the remainder of the study.

3.3. Bioreactor Culture Parameters: Agitation Speed, Day of Bioreactor Feed, Microcarrier Addition

We have shown that the addition of a bioreactor feed is necessary to obtain optimal cell growth (Figure 1A) without costly media exchanges (Figure 2A). With the fed-batch process in place, further optimization on the bioreactor culture process was performed. The increase of agitation speed on Day 3 was initiated due to the rapid cell growth following the feed addition, which often results in a high degree of cell aggregation (Figure 1C, Figure S3). Cell/microcarrier aggregation can be caused by cell-cell connections aided by the production of extracellular matrix (ECM) [40], which acts as an adhesive to bridge microcarriers upon collision during culture. This phenomenon started to be evident on Day 3 of the culture (Figure 1C). Subsequent increase in aggregate size may be due to increased cell number following the bioreactor feed addition and hence increased production of ECM.

Extensive cell aggregation has been mentioned to possibly correlate with the end of the exponential growth phase, as the cells within these aggregates were unable to receive sufficient nutrients and oxygen [41]. In fact, cell viability in hMSC aggregates has been shown to decrease as aggregate size increases due to limited nutrient transfer to the cells [40,42], therefore for optimal cell yield inside the bioreactor, aggregate size needs to remain small enough for it to continue being viable during culture.

To prevent cells from forming large aggregates, agitation speed can be varied. The cell/microcarrier suspension was subjected to two agitation speeds following addition of the bioreactor feed on Day 3: 30 rpm or 45 rpm. The bioreactor culture that is subjected to higher agitation speed showed slower growth after Day 3 (Figure 4A). A similar trend has been observed in other suspension bioreactor systems, where agitation above the optimal speed that keeps the cells in suspension results in a decrease in cell yield [43]. The reduction of cell growth at a higher agitation speed could be attributed to local shear-induced mechanical damage/cell detachment [8,9] or high levels of lactate production that inhibits cell growth [43].

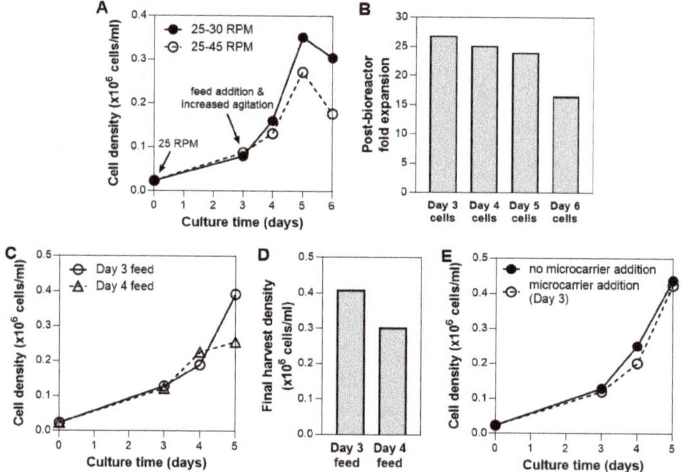

Figure 4. Optimization of bioreactor culture process. (**A**) Effect of bioreactor agitation speed on hMSC growth. Higher agitation speed (45 rpm) results in lower final cell yield. (**B**) Cell health following bioreactor culture. Cells harvested from the bioreactor on Day 6 showed 30% reduced expansion potential. (**C**,**D**) Effect of bioreactor feed timing on hMSC growth. Addition of feed on Day 3 results in the highest cell yield. (**E**) Effect of microcarrier addition during culture. Addition of 1.25 g of microcarriers (additional 450 cm^2 of surface area) to increase growth surface area during culture did not influence overall hMSC growth.

The cell growth kinetics also showed a decline in cell density after 5 days of culture (Figure 4A). In addition, when cells sampled from each day of bioreactor culture were plated, we observed >30% decline in expansion potential for cells harvested from Day 6 compared to Day 5 (Figure 4B). To both maximize cell yield and maintain post-harvest cell health, bioreactor harvest was therefore performed on Day 5 moving forward.

Another process parameter that is critical during a fed-batch bioreactor culture is the timing of the feed addition. Investigation of the effect of dynamic nutrient feeding in bioreactors has been performed in other cell types, where feedback control on nutrient demand was utilized to optimize for efficient cell metabolism [44,45]. Similarly, in our system, the effect of feeding time was evaluated. We observed that culture with feed on Day 4 resulted in a yield that was >25% lower than culture fed on Day 3 (Figure 4C,D). As seen in the shape of the growth kinetic curves, feed addition on Day 3 is necessary to maintain exponential cell growth over 5 days, while Day 4 addition was after media limitations had led to irreversible adverse effects on growth rates. In separate experiments, we also observed that addition of feed on Day 2 or the addition of two feeds (on Day 3 and 4, or Day 2 and 4), did not result in higher cell yield compared to feed addition on Day 3 (data not shown). This was consistent with our cytokine analysis data from the bioreactor spent media, which showed rapid depletion of FGF in bioreactor culture within days (Figure S4). The bioreactor feed was therefore added on Day 3 to replenish the media in subsequent studies.

Bead-to-bead cell transfer has been demonstrated in cultures of various cell types in various bioreactor systems [46–49]. The addition of fresh, empty microcarriers to an ongoing suspension culture could therefore potentially increase yield by providing additional area for cell growth (assuming media is not the limitation). For hMSCs, it has been reported that the bead-to-bead transfer process depends on the microcarrier substrate and culture medium [48]. In this system, the addition of 1.25 g of microcarriers (additional 450 cm^2 of surface area) on Day 3 of bioreactor culture showed no effect on cell growth (Figure 4E), possibly due to minimal cell migration between microcarriers. Complete investigation of bioprocess parameters that may allow the addition of microcarriers to in fact increase cell yield are beyond the scope of these studies.

3.4. Addition of Surfactant

While oxygen supply for hMSC culture at this scale (0.1 L) is sufficient through headspace gassing, sparging is often necessary to increase medium oxygenation to large-scale cultures with higher densities [50]. As bioreactors are operated at increasing working volumes, the ratio of the surface area available for headspace gassing to the culture volume decreases, therefore sparging could be implemented to meet the increasing oxygen demand [51]. The presence of gas bubbles from sparging, however, are known to cause cell damage and thus addition of a surfactant or antifoam are often implemented to reduce this damage [52,53]. We evaluated the addition of Pluronic F68 (PF68), a common surfactant used in cell culture to protect cell membranes from shear [54–56], at various concentrations and quantified the effect on cell growth in the bioreactor and the post-harvest cell health. Incorporation of PF68 on Day 3 at concentrations up to 2 g/L did not hinder cell growth (Figure 5A), however the use of 2 g/L of PF68 in the bioreactor culture reduced the cell expansion following bioreactor harvest by >50% compared to when 1 g/L PF68 was used (Figure 5B). Therefore, when sparging is necessary, a dosimetry study of pluronic concentration is necessary to determine the impact on post-expansion cell health, and in the case of PF68 a concentration of ≤1 g/L is recommended.

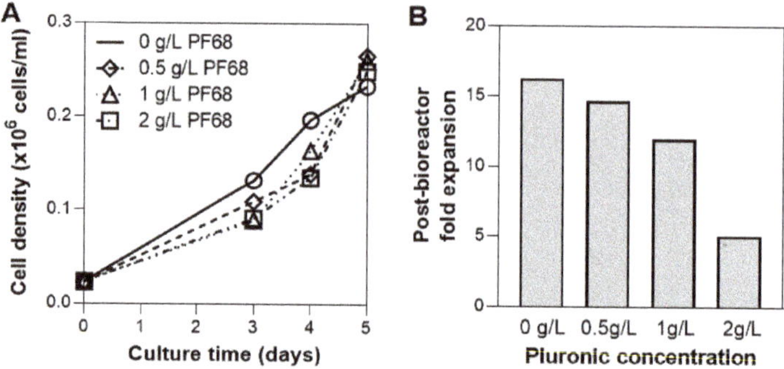

Figure 5. Effect of surfactants on hMSC growth in bioreactor. Addition of pluronic PF68 shows no effect on cell growth (**A**), however a high concentration of PF68 (2 g/L) markedly reduces post-bioreactor expansion (**B**).

3.5. Bioreactor Harvest Process: Agitation Speed, Quench Hold Time, Quench Solution Temperature

Following 5-day culture, hMSCs were harvested from the bioreactor. To date, the effect of harvest parameters on hMSC health/characteristics has been mainly focused on the choice of dissociation buffer and its effect on harvest yield and cell viability [57–59]. Here, we evaluated additional harvest process parameters, where agitation speed and duration were varied during the cell dissociation from the microcarriers using trypLE. The effect of dissociation duration in 2D cultures has been previously shown to affect MSC phenotype [59,60]. In our system, it was observed that high agitation (100 rpm) for 45 min resulted in the highest post-harvest cell health compared to multiple cycles at various agitation speeds (Figure 6A).

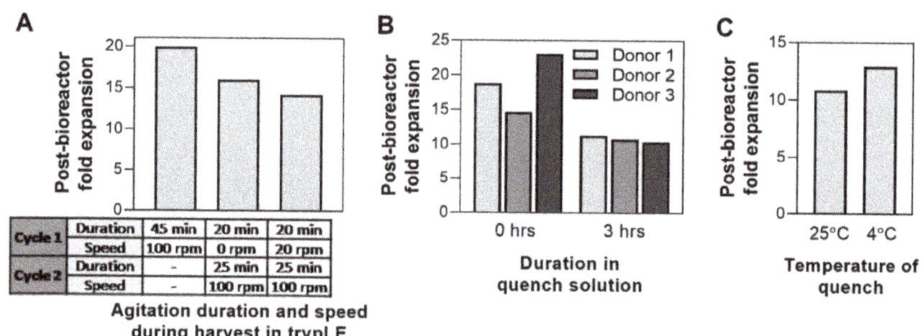

Figure 6. Effect of various bioreactor harvest conditions on post-harvest cell health: (**A**) agitation duration and speed during harvest in TrypLE, (**B**) duration in quench solution, and (**C**) temperature of quench solution.

Following cell detachment from the microcarriers, the enzymatic reaction was quenched by the addition of medium. Downstream processing of a large-scale bioreactor process, especially volume reduction and wash, can often take up to 4 h of processing time [61]; therefore, the effect of processing time on cell health was investigated. We compared the expansion of cells harvested from the bioreactor, where cells were plated straight after harvest (0 h) or after 3 h in quench solution. Increase of processing time to 3 h reduces the post-bioreactor cell expansion up to 54% in one donor (Figure 6B). Process development for bioreactor harvest and any further downstream processing should therefore be optimized to minimize time, particularly when scaling up to a production scale. Following harvest, the use of quench solution at 4 °C for 3 h resulted in higher post-bioreactor expansion compared to those using room temperature quench (25 °C) (Figure 6C). Although the difference is small (<20%), longer processing times likely amplify the disparity between processing at the two different temperatures. The use of a chiller is therefore recommended during downstream processing of cells from a large-scale bioreactor to optimize for cell health.

3.6. Validation of hMSC Expansion in Fed-batch Bioreactor Process and Maintenance of Critical Quality Attributes

Quantification of the effects of various bioreactor process parameters on the cell yield and post-bioreactor health resulted in an optimized protocol for hMSC expansion in a vertical-wheel suspension bioreactor (Figure 7A). This fed-batch protocol was validated with expansion of 5 hMSC donors in bioreactors (Figure 7B).

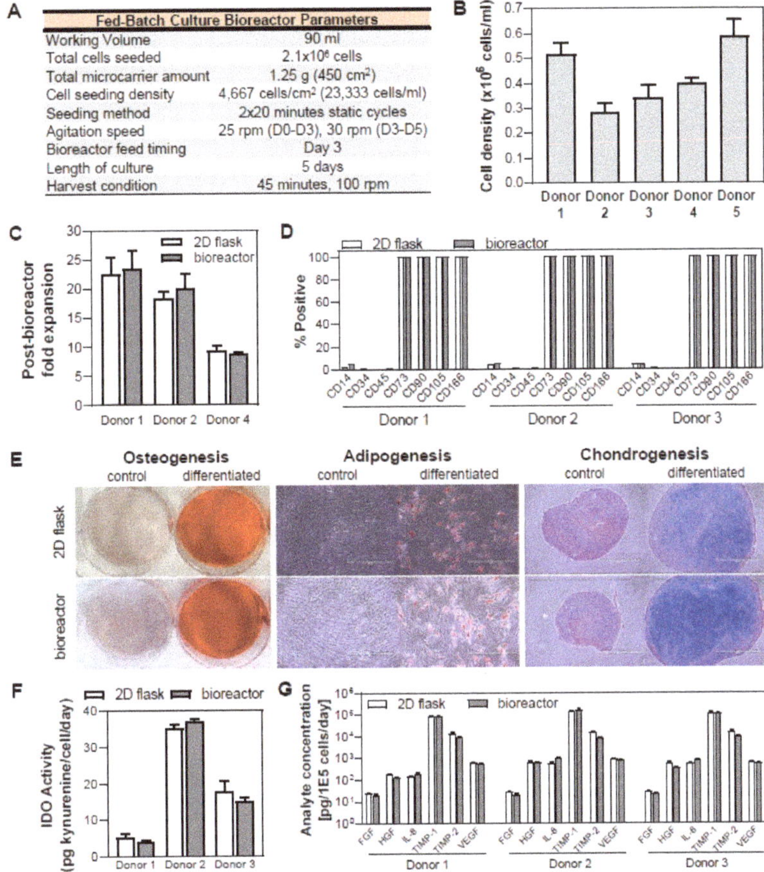

Figure 7. Validation of optimized fed-batch bioreactor process and hMSC characterization following bioreactor culture. (**A**) Parameters for fed-batch culture in a bioreactor. (**B**) Validation of optimized bioreactor culture process in five hMSC donors. (**C–G**) Verification of cell health, critical quality attributes, and functionality following bioreactor culture: XF hMSCs expanded in the bioreactor maintained their cell surface marker expression identity (**C**), tri-lineage differentiation potential to osteo-, adipo-, and chondrocytes (**E**), inducible indoleamine 2,3-dioxygenase (IDO) activity when stimulated with IFN$_\gamma$ (**F**), and cytokine secretion profile (**G**). The health, critical quality attributes, and functionality of three hMSC donors are comparable for cells harvested from bioreactor and cells harvested from its respective 2D flask control (In panel E, Scale bar = 400 μm).

Following bioreactor culture, it is important to verify the cell health, quality, and functionality. The cell health following bioreactor culture was verified by quantifying the subsequent 2D expansion of the cells harvested from the bioreactor. For three donors, this cell expansion at P5 was comparable to each donor's respective 2D control at similar population doubling level (PDL) (Figure 7C). In addition, the critical quality attributes (CQAs) of the cells need to be maintained and the functionality of the cells need to remain intact. Thus, we performed a general analytic panel to ensure CQAs of hMSCs, including cell identity (i.e., cell surface marker expression), trilineage differentiation potential, and functionality testing.

We observed that cells harvested from the bioreactor maintained their CQAs of hMSC surface marker expression of CD14, CD34, CD45, CD73, CD90, CD105 and CD166 (Figure 7D) as well as osteogenic, adipogenic, and chondrogenic differentiation potential (Figure 7E), therefore meeting ISCT's minimal criteria for defining MSCs [62]. In addition, functionality of cells from the bioreactor is also maintained, as indicated by inducible immunomodulatory potential (as measured by functional IDO activity) (Figure 7F) and angiogenic cytokine (FGF, HGF, IL-8, TIMP-1, TIMP-2, and VEGF) secretion (Figure 7G), where these values were comparable to those of cells grown in 2D culture with similar PDL. While the representative qualitative images for tri-lineage differentiation potential are shown for a single donor, images for three hMSC donors are presented in Figure S5.

Our developed bioreactor process yielded a range of harvest densities of 2×10^5–6×10^5 cells/mL, which was within the reported range of hMSC yield in various microcarrier-based bioreactor systems [63]. Previous studies of hBM-MSC expansion conducted in small scale (≤1 L) microcarrier suspension bioreactors have yielded 7.5×10^4–4.8×10^5 cells/mL density, with most of the studies being performed under serum-containing conditions in a 7–14 day bioreactor process [63]. In xeno-free cultures, cell densities of 2×10^5–3×10^5 cells/mL were achieved after 14 days [63]. In comparison, similar yields were achieved in these studies within 5 days of culture using a fed-batch process.

With this developed fed-batch microcarrier hMSC culture process, scaleup is the necessary next step to meet the demands for commercial manufacturing. In addition to verifying and further optimizing these parameters at larger scales, there are other considerations in large-scale production of hMSCs in bioreactors, such as the availability of platforms for downstream processing [61,64]. For example, while separation of microcarriers from the cells during the harvest step was done using a 100 μm cell strainer in our study, this method is not feasible for processes ≥1 L for which single-use filter bag technology is more appropriate. Furthermore, a cell concentration step is required prior to the final formulation and fill of the cell product. Large volume manufacturing of hMSCs demands scalable technologies for this step such as continuous counterflow centrifugation systems and newer technologies such as the acoustic wave cell processing, all of which are commercially available.

4. Conclusions

We developed a fed-batch, microcarrier-based bioreactor process, which enhances media productivity and drives a cost-effective and less labor-intensive hMSC expansion process. In this study, we determined parameters for various stages of the bioreactor culture: cell number and microcarrier concentration at inoculation, agitation speed during expansion, feed timing, culture length, as well as agitation speed and duration during the harvest process. We found that bioreactor inoculation at 5 cm^2 microcarriers surface area per mL of media and cell seeding density of 23,333 cells/mL resulted in an optimized yield and post-harvest cell health. The addition of a bioreactor feed on the third day of culture was critical to maintain the cell expansion by replenishing the mitogenic factors that were depleted from the medium. Our study resulted in an optimized hMSC culture protocol in a vertical-wheel suspension bioreactor. Our process is robust and works in multiple donors, and yields hMSCs with maintained expansion capability, phenotypic and functional properties that are comparable to traditional 2D flask cultures. The evaluated parameters were all investigated due to their relevance for larger scale production of hMSCs and parameter settings were selected both on biological response as well as system scalability to move towards production of hMSCs in larger bioreactors

(>80 L). We intend to continue to scale up this process to development, pilot, and production scale bioreactors, which are critical steps to generating the cell numbers necessary for clinical therapies.

Supplementary Materials: The following are available online at http://www.mdpi.com/2306-5354/7/3/73/s1, Figure S1: 24-hour cell attachment efficiency on microcarriers, Figure S2: Bioreactor-to-bioreactor variability within an experiment, Figure S3: hMSC cultures were sampled and monitored for cell growth on microcarriers during bioreactor culture, Figure S4: Depletion of FGF in bioreactor culture, Figure S5: Tri-lineage differentiation potential (to osteo-, adipo-, and chondrocytes) of 3 hMSC donors, comparing cells harvested from the bioreactor and each donor's respective 2D flask control.

Author Contributions: Conceptualization, J.L., R.K., L.T.L., J.A.R. and T.A.; methodology, J.L., R.K., J.D.T., T.R.O.; data generation and analysis, J.L., R.K., J.D.T., T.R.O.; writing—original draft preparation, J.L. and R.K.; writing—review and editing, J.L., R.K., J.A.R. and T.A. All authors have read and agreed to the published version of the manuscript.

Funding: This work was supported by the Medical Technology Enterprise Consortium (MTEC 16-01-REGEN-02).

Conflicts of Interest: The authors declare no conflicts of interest.

References

1. Olsen, T.R.; Ng, K.S.; Lock, L.T.; Ahsan, T.; Rowley, J.A. Peak MSC-Are We There Yet? *Front. Med. (Lausanne)* **2018**, *5*, 178. [CrossRef]
2. Rowley, J.A. Meeting Lot-Size Challenges of Manufacturing Adherent Cells for Therapy. *Bioprocess. Int.* **2012**, *10*, 16–22.
3. Simaria, A.S.; Hassan, S.; Varadaraju, H.; Rowley, J.; Warren, K.; Vanek, P.; Farid, S.S. Allogeneic cell therapy bioprocess economics and optimization: Single-use cell expansion technologies. *Biotechnol. Bioeng.* **2014**, *111*, 69–83. [CrossRef]
4. Chen, A.K.-L.; Chew, Y.K.; Tan, H.Y.; Reuveny, S.; Oh, S.K.W. Increasing efficiency of human mesenchymal stromal cell culture by optimization of microcarrier concentration and design of medium feed. *Cytotherapy* **2015**, *17*, 163–173. [CrossRef]
5. Schnitzler, A.C.; Kehoe, D.; Simler, J.; DiLeo, A.; Ball, A. Scale-up of Human Mesenchymal Stem Cells on Microcarriers in Suspension in a Single-use Bioreactor. *Bioprocess. Int.* **2012**, *25*, 28–39.
6. Szczypka, M.; Splan, D.; Woolls, H.; Brandwein, H. Single-Use Bioreactors and Microcarriers. *Bioprocess. Int.* **2014**, *12*, 54–64.
7. Yourek, G.; McCormick, S.M.; Mao, J.J.; Reilly, G.C. Shear stress induces osteogenic differentiation of human mesenchymal stem cells. *Regen. Med.* **2010**, *5*, 713–724. [CrossRef] [PubMed]
8. Dong, J.-D.; Gu, Y.-Q.; Li, C.-M.; Wang, C.-R.; Feng, Z.-G.; Qiu, R.-X.; Chen, B.; Li, J.-X.; Zhang, S.-W.; Wang, Z.-G.; et al. Response of mesenchymal stem cells to shear stress in tissue-engineered vascular grafts. *Acta Pharm. Sin.* **2009**, *30*, 530–536. [CrossRef] [PubMed]
9. Hung, B.P.; Babalola, O.M.; Bonassar, L.J. Quantitative characterization of mesenchymal stem cell adhesion to the articular cartilage surface. *J. Biomed. Mater. Res. A* **2013**, *101*, 3592–3598. [CrossRef] [PubMed]
10. Rafiq, Q.A.; Brosnan, K.M.; Coopman, K.; Nienow, A.W.; Hewitt, C. Culture of human mesenchymal stem cells on microcarriers in a 5 l stirred-tank bioreactor. *Biotechnol. Lett.* **2013**, *35*, 1233–1245. [CrossRef] [PubMed]
11. Lawson, T.; Kehoe, D.E.; Schnitzler, A.; Rapiejko, P.J.; Der, K.A.; Philbrick, K.; Punreddy, S.; Rigby, S.; Smith, R.; Feng, Q.; et al. Process development for expansion of human mesenchymal stromal cells in a 50 L single-use stirred tank bioreactor. *Biochem. Eng. J.* **2017**, *120*, 49–62. [CrossRef]
12. Kehoe, D.; Niss, K.; Rook, M.; Punreddy, S.; Murrel, J.; Sunil, N.; Aysola, M.; Jing, D. Growth Kinetics of Human Mesenchymal Stem Cells in a 3-L Single-Use, Stirred-Tank Bioreactor. *Biopharm. Int.* **2013**, *26*, 28–38.
13. Mizukami, A.; Fernandes-Platzgummer, A.; Carmelo, J.G.; Swiech, K.; Covas, D.T.; Cabral, J.M.S.; da Silva, C.L. Stirred tank bioreactor culture combined with serum-/xenogeneic-free culture medium enables an efficient expansion of umbilical cord-derived mesenchymal stem/stromal cells. *Biotechnol. J.* **2016**, *11*, 1048–1059. [CrossRef] [PubMed]
14. Croughan, M.; Giroux, D.; Fang, D.; Lee, B. Novel single-use bioreactors for scale-up of Anchorage-dependent cell manufacturing for cell therapies. In *Stem Cell Manufacturing*; Cabral, J.M.S., Chase, de Silva, C.L., Diogo, M.M., Eds.; Elsevier: Amsterdam, The Netherlands, 2016; pp. 105–139.

15. Cabral, J.M.S.; da Silva, C.L. *Bioreactors for Stem Cell Expansion and Differentiation*, 1st ed.; Atala, A., Almeida-Porada, G., Eds.; CRC Press: Boca Raton, FL, USA, 2018.
16. Pinto, D.D.S.; Bandeiras, C.; Fuzeta, M.D.A.; Rodrigues, C.A.V.; Jung, S.; Hashimura, Y.; Tseng, R.; Milligan, W.; Lee, B.; Ferreira, F.C.; et al. Scalable Manufacturing of Human Mesenchymal Stromal Cells in the Vertical-Wheel Bioreactor System: An Experimental and Economic Approach. *Biotechnol. J.* **2019**, *14*, 1800716. [CrossRef] [PubMed]
17. Lee, B.; Giroux, D.; Hashimura, Y.; Starkweather, N.; Rosello, F.; Wesselschmidt, R.; Croughan, M. New Scalable Manufacturing Platform for Shear-Sensitive Cell Therapy Products. *Cytotherapy* **2016**, *18*, S140. [CrossRef]
18. Sousa, M.; Silva, M.M.; Giroux, D.; Hashimura, Y.; Wesselschmidt, R.; Lee, B.; Roldão, A.; Carrondo, M.J.T.; Alves, P.M.; Serra, M. Production of oncolytic adenovirus and human mesenchymal stem cells in a single-use, Vertical-Wheel bioreactor system: Impact of bioreactor design on performance of microcarrier-based cell culture processes. *Biotechnol. Prog.* **2015**, *31*, 1600–1612. [CrossRef]
19. Shahdadfar, A.; Frønsdal, K.; Haug, T.; Reinholt, F.P.; Brinchmann, J.E. In vitro expansion of human mesenchymal stem cells: Choice of serum is a determinant of cell proliferation, differentiation, gene expression, and transcriptome stability. *Stem Cells* **2005**, *23*, 1357–1366. [CrossRef]
20. Cobo, F.; Stacey, G.N.; Hunt, C.; Cabrera, C.; Nieto, A.; Montes, R.; Cortés, J.L.; Catalina, P.; Barnie, A.; Concha, Á. Microbiological control in stem cell banks: Approaches to standardisation. *Appl. Microbiol. Biotechnol.* **2005**, *68*, 456–466. [CrossRef] [PubMed]
21. Wang, Y.; Han, Z.-B.; Song, Y.-P.; Han, Z.C. Safety of mesenchymal stem cells for clinical application. *Stem Cells Int.* **2012**, *2012*, 652034. [CrossRef] [PubMed]
22. Stute, N.; Holtz, K.; Bubenheim, M.; Lange, C.; Blake, F.; Zander, A.R. Autologous serum for isolation and expansion of human mesenchymal stem cells for clinical use. *Exp. Hematol.* **2004**, *32*, 1212–1225. [CrossRef] [PubMed]
23. Capelli, C.; Domenghini, M.; Borleri, G.; Bellavita, P.; Poma, R.; Carobbio, A.; Micò, C.; Rambaldi, A.; Golay, J.; Introna, M.; et al. Human platelet lysate allows expansion and clinical grade production of mesenchymal stromal cells from small samples of bone marrow aspirates or marrow filter washouts. *Bone Marrow Transpl.* **2007**, *40*, 785–791. [CrossRef] [PubMed]
24. Jung, S.; Sen, A.; Rosenberg, L.; Behie, L.A. Identification of growth and attachment factors for the serum-free isolation and expansion of human mesenchymal stromal cells. *Cytotherapy* **2010**, *12*, 637–657. [CrossRef]
25. Goh, T.K.-P.; Zhang, Z.; Chen, A.K.-L.; Reuveny, S.; Choolani, M.; Chan, J.; Oh, S.K.-W. Microcarrier culture for efficient expansion and osteogenic differentiation of human fetal mesenchymal stem cells. *BioRes. Open Access* **2013**, *2*, 84–97. [CrossRef] [PubMed]
26. Frauenschuh, S.; Reichmann, E.; Ibold, Y.; Goetz, P.M.; Sittinger, M.; Ringe, J. A microcarrier-based cultivation system for expansion of primary mesenchymal stem cells. *Biotechnol. Prog.* **2007**, *23*, 187–193. [CrossRef]
27. Santos, F.; Andrade, P.Z.; Abecasis, M.M.; Gimble, J.M.; Chase, L.G.; Campbell, A.M.; Boucher, S.; Vemuri, M.C.; Da Silva, C.L.; Cabral, J.M.S. Toward a clinical-grade expansion of mesenchymal stem cells from human sources: A microcarrier-based culture system under xeno-free conditions. *Tissue Eng. Part C Methods* **2011**, *17*, 1201–1210. [CrossRef] [PubMed]
28. Heathman, T.R.J.; Glyn, V.A.M.; Picken, A.; Rafiq, Q.A.; Coopman, K.; Nienow, A.W.; Kara, B.; Hewitt, C. Expansion, harvest and cryopreservation of human mesenchymal stem cells in a serum-free microcarrier process. *Biotechnol. Bioeng.* **2015**, *112*, 1696–1707. [CrossRef]
29. Nienow, A.; Rafiq, Q.A.; Coopman, K.; Hewitt, C. A potentially scalable method for the harvesting of hMSCs from microcarriers. *Biochem. Eng. J.* **2014**, *85*, 79–88. [CrossRef]
30. Rowley, J.A.; Montgomery, S.A. The Need for Adherent Cell Manufacturing: Production Platform and Media Strategies Drive Cell Production Economics. *BioProcess Int.* **2018**, *16*, 34–49.
31. The Glossary for Cell & Gene Therapy and Regenerative Medicine. *Regen. Med.* **2018**, *13*, 1–124.
32. Gjoka, X.; Gantier, R.; Schofield, M. Going from Fed-Batch to Perfusion. *Biopharm. Int.* **2017**, *30*, 32.
33. Abraham, E.; Gupta, S.; Jung, S.; McAfee, E. *Bioreactor for Scale-Up: Process Control, in Mesenchymal Stromal Cells: Translational Pathways to Clinical Adoption*; Viswanathan, S., Hematti, P., Eds.; Elsevier: Amsterdam, The Netherlands, 2016.
34. Krampera, M.; Galipeau, J.; Shi, Y.; Tarte, K.; Sensebé, L. Immunological characterization of multipotent mesenchymal stromal cells—The International Society for Cellular Therapy (ISCT) working proposal. *Cytotherapy* **2013**, *15*, 1054–1061. [CrossRef] [PubMed]

35. François, M.; Galipeau, J. New insights on translational development of mesenchymal stromal cells for suppressor therapy. *J. Cell. Physiol.* **2012**, *227*, 3535–3538. [CrossRef] [PubMed]
36. Meisel, R.P.; Zibert, A.; Laryea, M.; Gobel, U.; Daubener, W.; Dilloo, D. Human bone marrow stromal cells inhibit allogeneic T-cell responses by indoleamine 2,3-dioxygenase-mediated tryptophan degradation. *Blood* **2004**, *103*, 4619–4621. [CrossRef]
37. Boomsma, R.A.; Geenen, D.L. Mesenchymal stem cells secrete multiple cytokines that promote angiogenesis and have contrasting effects on chemotaxis and apoptosis. *PLoS ONE* **2012**, *7*, e35685. [CrossRef]
38. Duffy, G.; Ahsan, T.; O'Brien, T.; Barry, F.; Nerem, R.M. Bone marrow-derived mesenchymal stem cells promote angiogenic processes in a time- and dose-dependent manner in vitro. *Tissue Eng. Part A* **2009**, *15*, 2459–2470. [CrossRef]
39. Kwon, H.M.; Hur, S.-M.; Park, K.-Y.; Kim, C.; Kim, Y.-M.; Kim, H.-S.; Shin, H.-C.; Won, M.H.; Ha, K.-S.; Kwon, Y.-G.; et al. Multiple paracrine factors secreted by mesenchymal stem cells contribute to angiogenesis. *Vasc. Pharm.* **2014**, *63*, 19–28. [CrossRef]
40. Ferrari, C.; Balandras, F.; Guedon, E.; Olmos, E.; Chevalot, I.; Marc, A. Limiting cell aggregation during mesenchymal stem cell expansion on microcarriers. *Biotechnol. Prog.* **2012**, *28*, 780–787. [CrossRef] [PubMed]
41. Caruso, S.R.; Orellana, M.D.; Mizukami, A.; Fernandes, T.R.; Fontes, A.; Suazo, C.A.T.; Oliveira, V.D.C.; Covas, D.T.; Swiech, K. Growth and functional harvesting of human mesenchymal stromal cells cultured on a microcarrier-based system. *Biotechnol. Prog.* **2014**, *30*, 889–895. [CrossRef]
42. Bartosh, T.J.; Ylostalo, J.; Mohammadipoor, A.; Bazhanov, N.; Coble, K.; Claypool, K.; Lee, R.H.; Choi, H.; Prockop, D.J. Aggregation of human mesenchymal stromal cells (MSCs) into 3D spheroids enhances their antiinflammatory properties. *Proc. Natl. Acad. Sci. USA* **2010**, *107*, 13724–13729. [CrossRef] [PubMed]
43. Jossen, V.; Schirmer, C.; Sindi, D.M.; Eibl, R.; Kraume, M.; Pörtner, R.; Eibl, D. Theoretical and Practical Issues That Are Relevant When Scaling Up hMSC Microcarrier Production Processes. *Stem Cells Int.* **2016**, *2016*, 4760414. [CrossRef]
44. Wong, D.C.F.; Wong, K.T.K.; Goh, L.T.; Heng, C.-K.; Yap, M.G.S. Impact of dynamic online fed-batch strategies on metabolism, productivity and N-glycosylation quality in CHO cell cultures. *Biotechnol. Bioeng.* **2005**, *89*, 164–177. [CrossRef] [PubMed]
45. Ofiteru, I.D.; Lavric, V.; Woinaroschy, A. A sensitivity analysis of the fed-batch animal-cell bioreactor with respect to some control parameters. *Biotechnol. Appl. Biochem.* **2005**, *41 Pt 1*, 29–35. [CrossRef]
46. Ohlson, S.; Branscomb, J.; Nilsson, K. Bead-to-bead transfer of Chinese hamster ovary cells using macroporous microcarriers. *Cytotechnology* **1994**, *14*, 67–80. [CrossRef] [PubMed]
47. Wang, Y.; Ouyang, F. Bead-to-bead transfer of Vero cells in microcarrier culture. *Cytotechnology* **1999**, *31*, 221–224. [CrossRef] [PubMed]
48. Leber, J.; Barekzai, J.; Blumenstock, M.; Pospisil, B.; Salzig, D.; Czermak, P. Microcarrier choice and bead-to-bead transfer for human mesenchymal stem cells in serum-containing and chemically defined media. *Process. Biochem.* **2017**, *59*, 255–265. [CrossRef]
49. Rafiq, Q.A.; Ruck, S.; Hanga, M.P.; Heathman, T.R.; Coopman, K.; Nienow, A.W.; Williams, D.J.; Hewitt, C. Qualitative and quantitative demonstration of bead-to-bead transfer with bone marrow-derived human mesenchymal stem cells on microcarriers: Utilising the phenomenon to improve culture performance. *Biochem. Eng. J.* **2018**, *135*, 11–21. [CrossRef]
50. Panchalingam, K.M.; Jung, S.; Rosenberg, L.; Behie, L.A. Bioprocessing strategies for the large-scale production of human mesenchymal stem cells: A review. *Stem Cell Res. Ther.* **2015**, *6*, 225–265. [CrossRef]
51. Nienow, A.W.; Coopman, K.; Heathman, T.R.J.; Rafiq, Q.A.; Hewitt, C.J. Bioreactor Engineering Fundamentals for Stem Cell Manufacturing. In *Stem Cell Manufacturing*; Cabral, J.M.S., de Silva, C.L., Chase, L.G., Diogo, M.M., Eds.; Elsevier: Amsterdam, The Netherlands, 2016; pp. 43–75.
52. Chisti, Y. Animal-cell damage in sparged bioreactors. *Trends Biotechnol.* **2000**, *18*, 420–432. [CrossRef]
53. Zhang, S.; Handa-Corrigan, A.; Spier, R. Foaming and media surfactant effects on the cultivation of animal cells in stirred and sparged bioreactors. *J. Biotechnol.* **1992**, *25*, 289–306. [CrossRef]
54. Leung, H.W.; Chen, A.; Choo, A.; Reuveny, S.; Oh, S.K.-W. Agitation can induce differentiation of human pluripotent stem cells in microcarrier cultures. *Tissue Eng. Part C Methods* **2011**, *17*, 165–172. [CrossRef]
55. Lecina, M.; Ting, S.; Choo, A.; Reuveny, S.; Oh, S. Scalable platform for human embryonic stem cell differentiation to cardiomyocytes in suspended microcarrier cultures. *Tissue Eng. Part C Methods* **2010**, *16*, 1609–1619. [CrossRef] [PubMed]

56. Hewitt, C.; Lee, K.; Nienow, A.W.; Thomas, R.; Smith, M.; Thomas, C.R. Expansion of human mesenchymal stem cells on microcarriers. *Biotechnol. Lett.* **2011**, *33*, 2325–2335. [CrossRef] [PubMed]
57. Weber, C.; Pohl, S.; Pörtner, R.; Wallrapp, C.; Kassem, M.; Geigle, P.; Czermak, P. Expansion and Harvesting of hMSC-TERT. *Open Biomed. Eng. J.* **2007**, *1*, 38–46. [CrossRef]
58. Heng, B.C.; Cowan, C.M.; Basu, S. Comparison of enzymatic and non-enzymatic means of dissociating adherent monolayers of mesenchymal stem cells. *Biol. Proced. Online* **2009**, *11*, 161–169. [CrossRef]
59. Tsuji, K.; Ojima, M.; Otabe, K.; Horie, M.; Koga, H.; Sekiya, I.; Muneta, T. Effects of Different Cell-Detaching Methods on the Viability and Cell Surface Antigen Expression of Synovial Mesenchymal Stem Cells. *Cell Transpl.* **2017**, *26*, 1089–1102. [CrossRef] [PubMed]
60. Potapova, I.A.; Brink, P.R.; Cohen, I.S.; Doronin, S.V. Culturing of human mesenchymal stem cells as three-dimensional aggregates induces functional expression of CXCR4 that regulates adhesion to endothelial cells. *J. Biol. Chem.* **2008**, *283*, 13100–13107. [CrossRef]
61. Pattasseril, J.; Varadaraju, H.; Lock, L.; Rowley, J.A. Downstream technology landscape for large-scale therapeutic cell processing. *Bioprocess. Int.* **2013**, *11*, 38–47.
62. Dominici, M.; Le Blanc, K.; Mueller, I.; Slaper-Cortenbach, I.; Marini, F.; Krause, D.; Deans, R.; Keating, A.; Prockop, D.; Horwitz, E. Minimal criteria for defining multipotent mesenchymal stromal cells. The International Society for Cellular Therapy position statement. *Cytotherapy* **2006**, *8*, 315–317. [CrossRef] [PubMed]
63. Tsai, A.-C.; Jeske, R.; Chen, X.; Yuan, X.; Li, Y. Influence of Microenvironment on Mesenchymal Stem Cell Therapeutic Potency: From Planar Culture to Microcarriers. *Front. Bioeng. Biotechnol.* **2020**, *8*, 640. [CrossRef]
64. Rowley, J. Developing Cell Therapy Biomanufacturing Processes. *Chem. Eng. Process.* **2010**, *106*, 50–55.

Article

Ex-Vivo Stimulation of Adipose Stem Cells by Growth Factors and Fibrin-Hydrogel Assisted Delivery Strategies for Treating Nerve Gap-Injuries

Katharina M. Prautsch [1,2,3], Lucas Degrugillier [1,2,3], Dirk J. Schaefer [1,4], Raphael Guzman [4,5], Daniel F. Kalbermatten [1,2,3,†] and Srinivas Madduri [1,2,3,4,*,†]

1 Department of Plastic, Reconstructive, Aesthetic and Hand Surgery, University Hospital Basel, University of Basel, Spitalstrasse 21, 4021 Basel, Switzerland; katharina.prautsch@unibas.ch (K.M.P.); LucasMichelJean.Degrugillier@usb.ch (L.D.); Dirk.Schaefer@usb.ch (D.J.S.); Daniel.Kalbermatten@usb.ch (D.F.K.)
2 Department of Pathology, University Hospital Basel, Hebelstrasse 20, 4021 Basel, Switzerland
3 Department of Biomedical Engineering, University of Basel, Gewerbestrasse 14, 4123 Allschwil, Switzerland
4 Department of Biomedicine, University Hospital Basel, Hebelstrasse 20, 4021 Basel, Switzerland
5 Department of Neurosurgery, University Hospital Basel, Spitalstrasse 21, 4021 Basel, Switzerland; Raphael.Guzman@usb.ch
* Correspondence: srinivas.madduri@usb.ch
† Daniel F. Kalbermatten and Srinivas Madduri share the senior authorship.

Received: 9 April 2020; Accepted: 3 May 2020; Published: 5 May 2020

Abstract: Peripheral nerve injuries often result in lifelong disabilities despite advanced surgical interventions, indicating the urgent clinical need for effective therapies. In order to improve the potency of adipose-derived stem cells (ASC) for nerve regeneration, the present study focused primarily on ex-vivo stimulation of ASC by using growth factors, i.e., nerve growth factor (NGF) or vascular endothelial growth factor (VEGF) and secondly on fibrin-hydrogel nerve conduits (FNC) assisted ASC delivery strategies, i.e., intramural vs. intraluminal loading. ASC were stimulated by NGF or VEGF for 3 days and the resulting secretome was subsequently evaluated in an in vitro axonal outgrowth assay. For the animal study, a 10 mm sciatic nerve gap-injury was created in rats and reconstructed using FNC loaded with ASC. Secretome derived from NGF-stimulated ASC promoted significant axonal outgrowth from the DRG-explants in comparison to all other conditions. Thus, NGF-stimulated ASC were further investigated in animals and found to enhance early nerve regeneration as evidenced by the increased number of β-Tubulin III+ axons. Notably, FNC assisted intramural delivery enabled the improvement of ASC's therapeutic efficacy in comparison to the intraluminal delivery system. Thus, ex-vivo stimulation of ASC by NGF and FNC assisted intramural delivery may offer new options for developing effective therapies.

Keywords: adipose stem cells; neurotrophic factors; growth factors; peripheral nerve injuries; fibrin nerve conduits; hydrogels; stem cells delivery; axonal regeneration; Schwann cells

1. Introduction

Peripheral nerve injuries often result in loss of sensory and motor functions due to lack of effective therapeutic strategies, thus there is a great clinical need for developing new therapies [1,2]. Schwann cells (SC) play a crucial role in neuronal survival, axonal regeneration and re-myelination [3,4] by secreting an array of molecular signals, neurotrophic factors (NTF) and extracellular matrix proteins [5]. Furthermore, SC generate bands of Büngners for topographical guidance and path finding of the regenerating axons [6]. However, therapeutic use of SC is hampered due to the problems associated with harvesting the cells from healthy nerves and resulting co-morbidities [7]. Therefore, the need for stem cell-based therapies emerged for treating nerve injuries.

Therapeutic stem cells should be easily accessible and undergo rapid proliferation and differentiation in vitro under controlled conditions. Mesenchymal stem cells (MSC) can differentiate into SC-like cells under specific stimuli and enhance peripheral nerve regeneration [8–11]. Not limiting to the source of bone marrow, MSC with multi lineage capacity can also be obtained from adipose tissue, dental pulp, umbilical cord blood and Wharton's jelly of umbilical cord [12–15]. Adipose-derived stem cells (ASC) are easily accessible in large quantities from fat-tissue enabled by liposuction or abdominoplasty [6,13,16]. ASC proliferate rapidly [6] and possess a multi-lineage capacity, i.e., adipocytes, osteoblasts, chondrocytes [13,17,18]. Furthermore, ASC retain their mesenchymal potency over long-term culture [6,18] and promote the nerve regeneration similar to bone marrow stem cells [7]. In allogeneic transplantation, usage of ASC, like any other adult stem cells, benefits from their hypo-immunogenicity or immune-privilege by virtue of the reduced expression of HLA-DR class II histocompatibility antigen [19,20]. Thus, ASC may enable the development of "off-the-shelf therapies" for promoting the nerve regeneration. Within this context, neurotrophic potency of ASC and their delivery approaches play a crucial role in improving therapeutic efficacy.

As shown by many studies, factors that are present in the secretome of ASC promote axonal regeneration both in vitro and in vivo. When co-cultured with motor neuron-like cells or dorsal root ganglion (DRG) sensory neurons, ASC enhanced the neurite number and outgrowth length [21–23]. In line with these findings, several animal studies reported enhanced axonal regeneration and sensory-motor nerve conduction, when ASC were transplanted in polymeric nerve conduits (NC) or fibrin-hydrogel nerve conduits (FNC) [24–26]. Improved axonal growth and elongation was found in the FNC loaded with ASC in comparison to empty FNC [24,27,28]. However, an optimal delivery route for ASC still remains to be established in the context of peripheral nerve reconstruction. Various studies involving ASC therapy employed different cell delivery routes, i.e., injection into the lumen, dispersion within the fibrin conduit wall, dispersion within the lumen of fibrin matrix and coating on the luminal surface [7,22,26,27,29,30]. However, the therapeutic efficacy of the cells was scarcely correlated with delivery routes and carrier matrix in the context of nerve regeneration. Thus, the ineffective and incomplete outcome achieved so far by using the ASC therapy can be attributed largely to various structural and biochemical microenvironments that were inadequately orchestrated through different delivery routes. Moreover, studies focusing on the cell delivery strategies, for understanding the spatiotemporal influence of carrier matrix on the efficacy of therapeutic cells, are largely missing. Therefore, there is a clear need to investigate systematically the impact of important local delivery strategies on nerve regeneration and to establish effective options.

The most important soluble factors known to support the neuronal survival and axonal regeneration following traumatic nerve injury are NTF [4,31–33]. Nerve growth factor (NGF), amongst other NTF, specifically promotes survival and regeneration of sensory neurons by binding to the high-affinity trk-A receptors [34]. Vascular endothelial growth factor (VEGF) on the other hand is a potent angiogenic factor, which promotes proliferation of endothelial cells, formation and permeability of vascular structures [35]. Nevertheless, several studies demonstrated VEGF for having neurotrophic activities that were mediated either by flk-1 and flt-1 receptor binding or by enhanced vascularization [35–39].

Same as for ASC, various studies reported the improved nerve regeneration supported by exogenously administered NTF [31,34,40,41]. Despite these findings, the growth-promoting effects of ASC in response to a specific growth factors' stimulus largely remain elusive in the context of nerve regeneration. On the other hand, rapid upregulation of NGF and VEGF after traumatic nerve injury [4,33] clearly shows the clinical significance for exposing the therapeutic impact of ASC following growth factors' stimulation in the context of axonal regeneration.

We hypothesize improvement in the neurotrophic capacity of ASC in response to exogenous growth factors stimulation. Furthermore, a novel design of the fibrin hydrogel nerve conduits would facilitate discrete options for cell delivery and for enhancement of their therapeutic efficacy. Thus, the present study evaluated the capacity of ASC in response to specific stimuli of NGF or VEGF for promoting the axonal regeneration in vitro and in vivo. Furthermore, the influence of the stem cell delivery route, i.e., FNC assisted intramural vs. intraluminal ASC loading on early nerve regeneration was also investigated in rats by using a 10 mm sciatic nerve gap-injury model.

2. Materials and Methods

2.1. Isolation and Culture of Adipose Stem Cells (ASC)

All the in vitro studies were conducted in accordance with the local veterinary commission in Basel, Switzerland (No. 2925). Visceral adipose tissue was harvested from adult Sprague-Dawley rats and processed under sterile conditions as described earlier [21]. Briefly, the fat tissue was rinsed in 0.01 M phosphate-buffered solution (PBS), minced and resulting tissue was digested with 0.15% (w/v) Type I Collagenase (Gibco Life Technologies, Cat. No. 17100017) for 1 h at 37 °C and centrifuged for 5 min at 1500 rpm and 4 °C. The pellet was re-suspended in growth medium (GM) i.e., Dulbecco's Modified Eagle's Medium (DMEM, Gibco, Cat. No. 41965039) supplemented with 10% Foetal Bovine Serum (PAN-Biotech, EU-approved, Cat. No. P40-47500) and 1% Penicillin/Streptomycin (BioConcept, Cat. No. 4-01F00-H). Isolated ASC were seeded at a density of 3000 cells/cm^2 and expanded at 37 °C with 5% CO_2 in a humid atmosphere; GM was changed every 72 h. Cells were passaged using 0.25% Trypsin-EDTA (BioConcept, Cat. No. 5-51F00-H) at 90% confluence and resulting cells at passage 2 (P2) or 3 (P3) were used for the experiments.

2.2. ASC Characterization

Rat ASC (P2) were seeded on 24 well plates for characterization. Cells were fixed in 4% paraformaldehyde (PFA) at room temperature (RT) for 10 min and permeabilized and blocked in 1% normal goat serum (NGS) in PBS (i.e., dilution buffer) for 60 min at RT. ASCs were incubated overnight at 4 °C with the human mesenchymal stromal cell markers monoclonal mouse anti-CD44 (1:1000), monoclonal mouse anti-CD90 (1:200), monoclonal mouse anti-CD105 (1:200) and monoclonal rabbit anti-CD29 (1:100), and the hematopoietic marker polyclonal rabbit anti-CD45 (1:500) (Abcam, Cat. No. ab93758). Cells were then washed in PBS and incubated for 60 min at RT with the secondary antibody goat anti-mouse Alexa Fluor 488 (1:500, Abcam, Cat. No. ab150109) and goat anti-rabbit Alexa Fluor 488 (1:500, Abcam, Cat. No. ab150061) and Hoechst 33258 nuclear staining (1:1000, Sigma Aldrich, Cat. No. 94403). Subsequently, digital images were acquired at 20× magnification (numerical aperture 0.45) by using a Nikon Eclipse Ti2 fluorescent inverted microscope (Nikon Eclipse Ti2-E, -E/B, Nikon Corporation, Japan) and a Photometrics prime 95B 25 mm camera (Teledyne photometrics, Tucson, AZ, USA). The images were automatically stitched by the Nikon NIS-Elements AR image analysis software (NIS- Elements AR Analysis 5.11.00 64-bit, Laboratory Imaging, spol. s.r.o., Praha, Czech Republic). Furthermore, immunostaining images of ASC for various markers were analyzed quantitatively using ImageJ.

2.3. Isolation of Chicken Embryonic Dorsal Root Ganglions (DRG)

Fertilized chicken eggs were obtained from Gepro Geflügelzucht AG (Flawil, Switzerland). The eggs were shipped at ambient temperature and incubated at 37.8 ± 0.2 °C under 100% relative humidity for 10 days (E10). After incubation, the eggs were cleaned with 70% ethanol and opened under a laminar airflow cabinet to collect the embryos. E10 embryos were dissected following a standard dissection protocol under a stereomicroscope [34]. DRG-explants were collected from the lumbar part of the spine and transferred to GM for cell culture.

2.4. ASC Stimulation and Secretome Harvest

ASC were seeded on 24 well plates at a density of 13,000/cm^2 and 500 μL of GM was added with or without supplementation of exogenous growth factors, i.e., 10 ng/mL of recombinant human NGF or recombinant mice VEGF, as indicated in the experimental design. NGF (Cat. No. 256-GF) and VEGF (Cat. No. 493-MV) were obtained from R&D Systems (Minneapolis, MN, USA). Cells were cultured for 72 h and no growth medium was exchanged during culture to enrich the secretome. After 72 h of enrichment, the resulting conditioned medium enriched with secretome (STM) was collected and used for subsequent experiments.

2.5. Experimental Design In Vitro

As illustrated in Figure 1, ASC were stimulated by NGF (NGF-ASC) or VEGF (VEGF-ASC) or without growth factor (ASC) for 72 h and resulting secretome (STM), i.e., STM-NGF-ASC, STM-VEGF-ASC and STM-ASC was subsequently used for DRG assay for 48 h. When the ASC were not stimulated prior to DRG assay, STM derived from the ASC was either supplemented with NGF (STM-ASC+NGF) or VEGF (STM-ASC+VEGF) or no growth factors (STM-ASC) at the time of DRG seeding. As a control, culture conditions with NGF alone (NGF) or VEGF alone (VEGF) or without growth factors (no GF) were used. In all cases, growth factors were applied at 10 ng/mL based on preliminary experiments [34]. For the axonal outgrowth assay, DRG-explants were seeded at a density of one per well onto 24 well plates. Cultures were maintained in a humid atmosphere at 37 °C and 5% CO_2 for 48 h and images were captured at 5× and 10× magnification. In total, 3 independent experiments resulting in a total of 8 DRG-explant cultures for each experimental condition were performed.

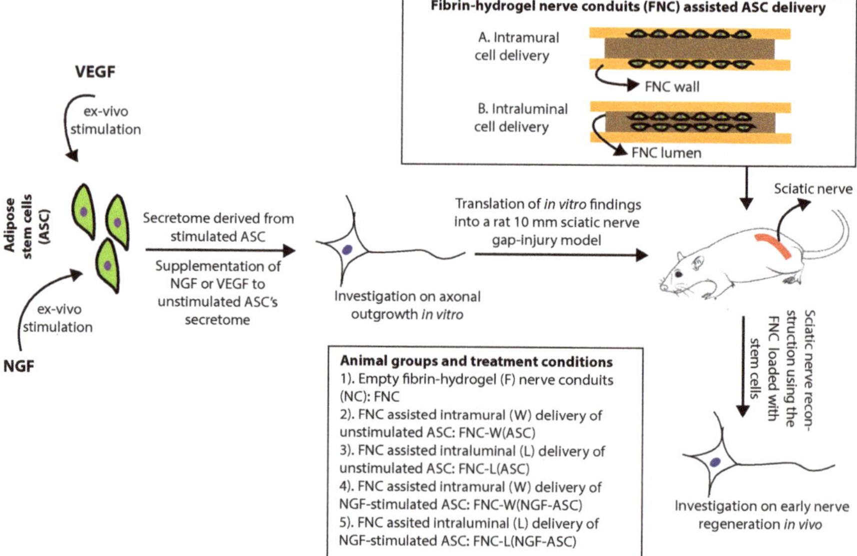

Figure 1. Ex-vivo stimulation of adipose stem cells using growth factors and fibrin-hydrogel nerve conduits assisted stem cell delivery strategies for enhancing the axonal regeneration.

2.6. Immunocytochemistry of DRG Cultures

After 48 h, DRG cultures were observed under a microscope and bright-field images with phase-contrast were taken at 5× magnification using a Zeiss Axio Vert.A1 inverted fluorescent microscope (Carl Zeiss AG, Jena, Germany) and a Zeiss AxioCam MRc camera (Carl Zeiss AG, Jena, Germany). DRG-explants were then fixed in 4% PFA at room temperature (RT) for 10 min and permeabilized and blocked in PBS containing 0.1% Triton X-100 and 1% BSA (i.e., dilution buffer) for 60 min at RT. For immunocytochemistry, the cultures were incubated overnight with the following primary antibody at 4 °C: monoclonal mouse anti-β-Tubulin III (1:1000, Sigma-Aldrich, Cat. No. T8578) for axons. The cultures were then washed in PBS and incubated with the following secondary antibody: sheep anti-mouse Cy3 (1:500, Sigma Aldrich, Cat. No. C2181) and Hoechst 33258 nuclear staining (1:1000, Sigma Aldrich, Cat. No. 94403) for 60 min at RT. Subsequently, digital images were acquired at 10× magnification (numerical aperture 0.45) by using a Nikon Eclipse Ti inverted fluorescent microscope and a Photometrics prime 95B 25 mm camera.

2.7. Quantitative Measurements of Axonal Outgrowth

DRG-explant cultures were analyzed for axonal length and axonal area in an automated manner with a standardized analysis mask created using the Nikon NIS-Elements AR image analysis software. The DRG area was defined in the Hoechst channel and the intensity threshold was set to be 45. The area occupied by axonal outgrowth was evaluated in the Cy3 channel and the intensity was set to be 65. For accurate axonal area measurements, no binary processing for "fill holes" was selected. For analysis of the axonal length, on the other hand, binary processing for "fill holes" was used. A binary operation expression was used for defining the origin of axonal outgrowth from DRG as well as axonal endpoints. The shortest perpendicular distances from the axonal growth origin to the endpoints were measured and the output was displayed automatically.

2.8. Stimulated ASC for Animal Studies

ASC were harvested and cultured as explained earlier. Briefly, ASC were cultured in 75 cm^2 flasks at a density of 13000 cells/cm^2 and stimulated by NGF 10 ng/mL for 72 h prior to transplantation.

2.9. Fibrin-Hydrogel Nerve Conduits (FNC)

FNC were prepared according to the manufacturer's instructions as described earlier (Tisseel Kit VH 1.0, Baxter, SA, USA) [42,43]. Briefly, Tisseel provides a fibrinogen solution containing fibrinogen 100 mg/mL, factor XIII 0.6–10 IU/mL, plasminogen 40–120 mg/mL, aprotinin synthetic 3000 KIU/mL, and a thrombin solution 500 IU/mL with calcium chloride 40 µmol/mL. Fibrin conduits measuring 14 mm in length, 1 mm in wall thickness and 2 mm in lumen were produced for bridging a 10-mm nerve gap injury. The thrombin 500 IU/mL was diluted in sterile water to 30 IU/mL. Equal amounts of the diluted thrombin and the fibrinogen solution were mixed directly into a silicone mold pre-set with a stainless steel lumen and incubated for 30 min at 37 °C for polymerization. Resulting FNC constructs were stored in sterile PBS at 4 °C for a maximum of 24 h before animal experiments.

2.10. Intramural Delivery of ASC

About 2 million NGF-stimulated ASC or unstimulated ASC were incorporated into the wall of the conduit, the resulting fibrin conduits were designated as FNC-W(NGF-ASC) and FNC-W(ASC) respectively. Immediately before transplantation, cells were trypsinized and suspended in a volume of 540 µL Fibrinogen. Further, 60 µL of thrombin solution 30 IU/mL was added to the ASC-Fibrinogen suspension and the resulting solution was cast into the FNC.

2.11. Intraluminal Delivery of ASC

About 2 million NGF-stimulated ASC or unstimulated ASC were incorporated into the lumen of the conduit, the resulting fibrin conduits were designated as FNC-L(NGF-ASC) and FNC-L(ASC) respectively. Cells were suspended in the final volume of 20 µL of Fibrinogen solution (25 mg/mL). Further, 20 µL of thrombin was added to the ASC-Fibrinogen suspension. The resulting solution was injected into the lumen of the FNC and incubated for 30 min at 37 °C for polymerization.

2.12. Surgical Procedure and Experimental Groups In Vivo

All the studies were conducted in accordance with the local veterinary commission in Basel, Switzerland (No. 2925). About 12-week old female Sprague-Dawley rats weighing 250–300 g (Janvier, Mayenne, France) were used for the experiment. A total of 36 rats were operated and randomly categorized into 6 groups: (1) autograft, (2) fibrin-hydrogel nerve conduit without cells, i.e., FNC, (3) FNC assisted intramural delivery of unstimulated ASC, i.e., FNC-W(ASC), (4) FNC assisted intramural delivery of NGF-stimulated ASC, i.e., FNC-W(NGF-ASC), (5) FNC assisted intraluminal delivery of unstimulated ASC, i.e., FNC-L(ASC), (6) FNC assisted intraluminal delivery of NGF-stimulated ASC, i.e., FNC-L(NGF-ASC). All surgical procedures were performed under general anesthesia with 3% isoflurane. Routinely, Buprenorphine (Temgesic ®, 0.3 mg/mL, 0.05 mg/kg, Indivior AG, Baar, Switzerland) was administered before and after the surgery. The left sciatic nerve was approached dorsally using a gentle spreading technique of the gluteus muscle. The sciatic nerve was transected 5 mm distal to the gluteal branch and nerve ends were inserted 2 mm into the fibrin conduit and fixed to the conduit by a single epineural suture (9/0 nylon, S&T). Muscles and fascia layers were closed with a single resorbable stitch (5/0 Vicryl, Ethicon) and the skin was closed by a continuous running suture (5/0 Vicryl, Ethicon). For analgesia, the animals received subcutaneous injections of Meloxicam (Metacam ®, 2 mg/mL, Boehringer Ingelheim GmbH, Basel, Switzerland) 2 mg/kg BW postoperatively, and then 1 mg/kg BW every 24 h for two days. Further, Paracetamol (Becetamol ®, 100 mg/mL, Gebro Pharma AG, Liestal, Switzerland) 50mg/kg BW/day was added to the drinking water for 7 days.

2.13. Tissue Processing and Immunohistochemistry

At 4 weeks post-operation, all the animals were euthanized by CO_2 and regenerated nerve tissue was harvested with the proximal and distal nerve stumps. The harvested nerves were embedded in OCT freezing media (Tissue-Tek, Sakura, Japan) and flash-frozen in liquid nitrogen through 2-methylbutane (Sigma Aldrich, Cat. No. M32631). Nerve cross-sections were prepared by cryostat from the middle, distal and far distal part of the explanted nerve tissue as shown in Figure 2K. The middle sections were taken right in the middle of the explanted conduit, the distal sections at the distal suture point, and the far distal sections 5mm distal to the distal suture. Serial 5 µm thick tissue sections were prepared onto slides (Superfrost plus, Menzel-Gläser, Braunschweig, Germany) and stored at −80 °C. For staining, every second section was processed. First, the slides were fixed in 4% PFA for 10 min and washed in distilled water and then blocked using dilution buffer for 60 min at RT. Slides were then incubated overnight at 4 °C with the following primary antibodies: monoclonal mouse anti-β-Tubulin III (1:1000, Sigma Aldrich, Cat. No. T8578) and polyclonal rabbit anti-S100 (1:100, abcam, Cat. No. ab76729). After rinsing in PBS, secondary sheep anti-mouse antibodies Cy3 (1:500, Sigma Aldrich, Cat. No. C2181), donkey anti-rabbit Alexa Fluor 488 (1:500 abcam, Cat. No. ab150061) and Hoechst 33258 (1:1000, Sigma Aldrich, Cat. No. 94403) were applied for 60 min at RT. The slides were mounted using ProLong Gold (Invitrogen, Cat. No. P36930) and digital images were acquired using a Nikon Eclipse Ti2 inverted fluorescent microscope and a Photometrics prime 95B 25 mm camera at 10× (numerical aperture 0.45), 20× (numerical aperture 0.75) and 40× (numerical aperture 0.95) magnification.

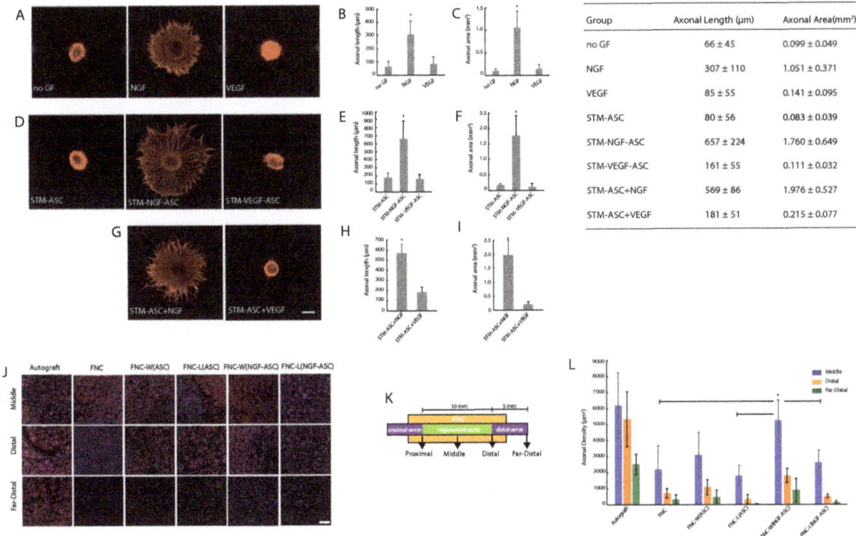

Group	Axonal Length (μm)	Axonal Area(mm²)
no GF	66 ± 45	0.099 ± 0.049
NGF	307 ± 110	1.051 ± 0.371
VEGF	85 ± 55	0.141 ± 0.095
STM-ASC	80 ± 56	0.083 ± 0.039
STM-NGF-ASC	657 ± 224	1.760 ± 0.649
STM-VEGF-ASC	161 ± 55	0.111 ± 0.032
STM-ASC+NGF	569 ± 86	1.976 ± 0.527
STM-ASC+VEGF	181 ± 51	0.215 ± 0.077

Figure 2. Fibrin assisted delivery of stimulated stem cells for axonal regeneration: (**A**) Microphotographs of DRG-explant cultures treated with growth factors. (**B**) Quantitative measurements of axonal length. (**C**) Quantitative measurements of the axonal area. (**D**) Microphotographs of DRG-explant cultures that were treated with secretome derived from unstimulated ASC (STM-ASC) or NGF-stimulated ASC (STM-NGF-ASC) or VEGF-stimulated ASC (STM-VEGF-ASC). (**E**) Quantitative measurements of axonal length. (**F**) Quantitative measurements of the axonal area. (**G**) Microphotographs of DRG-explant cultures that were treated with ASC's secretome in combination with exogenous NGF (STM-ASC+NGF) or VEGF (STM-ASC+VEGF). (**H**) Quantitative measurements of axonal length. (**I**) Quantitative measurements of the axonal area. For images (**A**)–(**I**), the scale bar represents 500 μm and the bars represent mean ± SD of n = 8. Significant differences at * $p < 0.05$ are indicated in comparison to all other treatment groups. (**J**) Microphotographs showing the β-Tubulin III+ axons for various experimental groups: Autograft; empty fibrin-hydrogel nerve conduit "FNC"; FNC's wall loaded with unstimulated ASC, i.e., intramural ASC delivery "FNC-W(ASC)"; FNC's lumen loaded with unstimulated ASC, i.e., intraluminal ASC delivery "FNC-L(ASC)"; FNC's wall loaded with NGF-stimulated ASC, i.e., intramural NGF-ASC delivery "FNC-W(NGF-ASC)"; FNC's lumen loaded with NGF-stimulated ASC, i.e., intraluminal NGF-ASC delivery "FNC-L(NGF-ASC)". The scale bar represents 100 μm. (**K**) Anatomical segmentation representing the histological analysis of the reconstructed nerves. (**L**) Quantitative measurements of β-Tubulin III+ axons for various experimental groups: Autograft; empty fibrin-hydrogel nerve conduit "FNC"; FNC's wall loaded with unstimulated ASC, i.e., intramural ASC delivery "FNC-W(ASC)"; FNC's lumen loaded with unstimulated ASC, i.e., intraluminal ASC delivery "FNC-L(ASC)"; FNC's wall loaded with NGF-stimulated ASC, i.e., intramural NGF-ASC delivery "FNC-W(NGF-ASC)"; FNC's lumen loaded with NGF-stimulated ASC, i.e., intraluminal NGF-ASC delivery "FNC-L(NGF-ASC)". Blue staining is Hoechst indicating cell nuclei. The bars represent mean ± SD of n = 6. Significant differences at * $p < 0.05$ are indicated in comparison to all other experimental groups.

2.14. Histological Analysis

Following immunofluorescence staining, digital images of 20× magnification were acquired and used for quantitative analysis of anatomical structures. For measuring the axonal density and area occupied by SC, an automated program was performed using the standardized analysis mask created by Nikon NIS-Elements AR image analysis software. Axonal count and nerve area values were used for calculation of axonal density. Similarly, the area occupied by SC was given in reference to the nerve area.

2.15. Statistical Analysis

Data were analyzed by two-way analysis of variance (ANOVA) following Bonferroni procedure with post hoc multiple comparisons using SPSS (version 15.0; SPSS, Chicago, IL, USA). Values with $p < 0.05$ were considered significant.

3. Results

3.1. Characterization of Isolated ASC

ASC were isolated, cultured and resulting cells were characterized phenotypically by immunocytochemistry. ASC were found to be positive for mesenchymal marker CD29 (87%), CD44 (78%), CD90 (81%) and CD105 (85%), and negative for hematopoietic marker CD45 (Figure S1).

3.2. Stem Cell Derived Secretome and Axonal Growth In Vitro

Consistent with our previous reports [34], DRG explants exhibited an important and dense axonal outgrowth in response to NGF-stimulation (Figure 2A). Quantitative measurements of axonal outgrowth, i.e., axonal length (in µm) and axonal area (in mm^2) resulted in 307 ± 110 and 1.05 ± 0.37 (Figure 2B,C). In contrast to NGF, stimulation with VEGF or without growth factors (no GF) resulted in only minimal axonal length, i.e., 85 ± 55 and 66 ± 45 (Figure 2B), which are consistent with axonal area measurements, i.e., 0.14 ± 0.10 and 0.10 ± 0.05 (Figure 2C) respectively.

Interestingly, STM-NGF-ASC enhanced the significant axonal outgrowth, i.e., 657 ± 224 and 1.76 ± 0.65 (Figure 2D,E). In the case of STM-ASC, no significant axonal outgrowth could be observed, i.e., 80 ± 56 and 0.083 ± 0.039 (Figure 2F). Together these observations clearly indicate the significantly enhanced potency of ASC in response to the NGF-stimulation for promoting axonal regeneration in vitro (Figure 2D–F).

In contrast to NGF conditions, STM-VEGF-ASC did not result in the enhancement of axonal outgrowth, i.e., 161 ± 55 and 0.111 ± 0.032 (Figure 2E). These observations indicate no significant improvement of ASC's potency in response to VEGF-stimulation for supporting axonal regeneration in vitro (Figure 2D–F).

In line with STM-NGF-ASC, STM-ASC+NGF culture condition resulted in a robust axonal outgrowth, i.e., 569 ± 86 and 1.98 ± 0.53 (Figure 2G–I). Together these results underline the important role of NGF for promoting axonal regeneration (Figure 2B,E,H). In contrast to these results, the STM-ASC+VEGF culture condition did not elicit synergy as evidenced by axonal outgrowth measurements, i.e., 181 ± 51 and 0.22 ± 0.077 (Figure 2H).

3.3. Stimulated Stem Cells and Delivery Route Impacted Early Nerve Regeneration

Considering in vitro findings on the enhanced potency of ASC in response to NGF-stimuli, NGF-ASC were further investigated in rats for treating a 10 mm sciatic nerve gap-injury. The impact of different ways of delivering these cells on the early nerve regeneration, i.e., intramural vs. intraluminal delivery, was also analyzed (Figure 1).

As depicted in Figure 2J, histological recovery of the treated animals was measured by analyzing the β-Tubulin III+ axons (Figure 2J,L) and S100 + SC structures (Supplementary Figure S2A,B) from the middle, distal and far-distal parts of the regenerated nerve tissue. In the middle part of the fibrin conduit, axonal regeneration (i.e., number of axons/ mm^2) resulted in the order of 6190 ± 2061, 5260 ± 1257, 3075 ± 1432, 2614 ± 743, 1810 ± 629 and 2197 ± 1478 for autograft, FNC-W(NGF-ASC), FNC-W(ASC), FNC-L(NGF-ASC), FNC-L(ASC) and FNC treatment groups respectively. These results reveal the enhanced axonal regeneration supported by the NGF-stimulated ASC. On the other hand, the superior performance of intramural stem cell delivery was evident compared to the intraluminal delivery route

(Figure 2J,L). Together, these results reveal the autograft matching performance of NGF-stimulated ASC in combination with the intramural delivery route i.e., FNC-W(NGF-ASC). Furthermore, the general tendency of enhanced axonal regeneration supported by the FNC-W(NGF-ASC) is evident in distal and far-distal segments in comparison to all other treatment groups (Figure 2J,L), although the differences between some of the groups were not statistically significant (Figure 2L).

In contrast to the distinct axonal growth response (Figure 2J,L) found for the various stimuli conditions, SC exhibited no differences (Figure S2A,B). However, the molecular differences for the recruited SC in the animals due to the various stimuli conditions remain to be investigated. We hypothesize that the Schwann cells recruited in response to the various stimuli conditions may vary in their phenotype (sensory vs. motor) and secretome.

4. Discussion

We hypothesized impactful change in the ability of ASC in response to growth factors' stimuli for supporting the axonal regeneration. Therefore, the present study was designed to investigate the growth promoting capacity of ASC following NGF- or VEGF-stimulation in vitro and to further translate in vitro findings into animals in combination with two different cell delivery approaches, i.e., FNC assisted intramural vs. intraluminal ASC loading for studying the early nerve regeneration.

In the present study, experiments were conducted using VEGF or NGF stimulated-ASC. A large portion of the research investigating the neurotrophic effect of ASC involved SCLC derived from ASC while studies on undifferentiated ASC remained scarce [5,44]. Although an SCLC phenotype seems to be desirable for optimal regenerative support, 2 to 4 weeks of in vitro differentiation process represents a major obstacle for the clinical use of differentiated ASC. On the other hand, the therapeutic benefits of differentiated vs. undifferentiated ASC are still not clear. Nevertheless, several studies suggest a similar therapeutic potential for both cell types [29,45,46]. There are two mechanisms explaining the regenerative potency of undifferentiated ASC. The first hypothesis states that ASC might undergo an in vivo trans-differentiation in response to the regenerative microenvironment [21,45,47,48]. The other hypothesis postulates that the neurotrophic potential of ASC may lie in the secretome containing a wide range of biochemical and molecular factors [49]. ASC exosomes releasing miRNA21, miRNA222 and miRNAlet7a play an important role in neuronal survival by inhibiting apoptotic pathways. In addition, the secretome of ASC contains nerve growth factor (NGF), glial cell-derived neurotrophic factor (GDNF), brain-derived neurotrophic factor (BDNF), neurotrophin-3 (NT-3), insulin-like growth factor 1 (IGF-1), vascular endothelial growth factor (VEGF), epidermal growth factor (EGF), basic fibroblast growth factor (bFGF), transforming growth factor beta (TGF-ß), and platelet-derived growth factors (PDGF) [50–54]. Furthermore, transplanted ASC may enhance the recruitment of endogenous SC to the injury site [29].

To the factors secreted by ASC, adult neurons may respond differently in contrast to embryonic neurons, as the trophic dependency of the various subsets of the neurons in the peripheral nervous system may change with the age. However, several studies have shown that injury-induced expression of various growth factors and surface receptors that are responsive for both adult and embryonic neurons [34,55]. Thus, the embryonic DRG assay used in our study should provide relevant information on the neurotrophic capacity of the secretome derived from ASC.

The data obtained in the present study clearly showed the enhanced ability of ASC in response to NGF-stimulus for axonal regeneration. The secretome derived from ASC after 72 h of NGF-stimulation (i.e., STM-NGF-ASC), as well as the secretome of unstimulated ASC supplemented with exogenous NGF (i.e., STM-ASC+NGF) induced the significant axonal outgrowth. Further on, there is no significant difference found in the axonal outgrowth promoted by either of these culture conditions indicating the growth-promoting function of NGF in combination with ASC's secretome. However, the exact mechanism of the underlying molecular functions needs to be determined along with the analysis of the secretome profile of ASC resulting from various experimental stimuli. On the contrary, VEGF culture conditions did not result in enhanced axonal regeneration in vitro. These observations may indicate no direct effect of VEGF in vitro on axonal regeneration. Interestingly, ASC secretome contains a wide range of growth factors, in addition to high amounts of VEGF (i.e., 2000 to 3000 pg/mL) [53,54] indicating that there may be no need for additional exogenous VEGF administration for ASC therapies. Thus, the in vitro results and data instructed the design of animal experiments using single factor NGF stimulated ASC.

In agreement with our in vitro findings, NGF-stimulated ASC exhibited a potential for promoting enhanced early nerve regeneration as evidenced by the increased number of β-Tubulin III+ axons. Functional restoration of human nerve injuries is critical and controlled by timely entry of regenerative axons into the distal nerve segment, which in turn is influenced by the quality (i.e., axonal growth speed) and quantity (axonal number) of early axonal regeneration. Thus, the present study reports on the important aspects of early nerve regeneration in the context of ASC-fibrin-hydrogel based therapeutic conduits.

In the middle of the fibrin conduit, i.e., regenerated nerve tissue, the density of regenerating axons for the animals treated with intramural delivery of NGF-stimulated ASC is statistically comparable to the autograft treatment, indicating not only the benefits of stimulated stem cells but also the importance of the cell delivery route. However, the outcome in the subsequent distal and far-distal segments is significantly higher for the autograft group. Further on intramural delivery of NGF-stimulated ASC, i.e., W(NGF-ASC) promoted better axonal regeneration as evidenced by the increased axonal number in the different parts of the conduit including the far-distal segment, although the axonal growth was found in the distal part of the conduit in all the other groups. The beneficial effects of delivering individual VEGF for nerve regeneration are unclear, given the number of studies reporting with variable outcomes, i.e., increased angiogenesis with or without functional benefits [56,57]. On the other hand, studies reporting in the literature with sequential release of VEGF and NGF exhibited beneficial effects. However, early axonal regeneration in those animals appeared poor than autograft animals as evidenced by significantly lower axonal regeneration in the middle graft after 4 weeks [58]. Therefore, the regeneration levels achieved in our present study appeared superior to that of dual-factor NGF and VEGF treatment, indicating the potential of the NGF stimulated-ASC for better nerve regeneration and for reducing the complexity of administering additional exogenous growth factors.

In order to improve the outcomes, i.e., suboptimal results obtained particularly at the distal part of the nerve, we would need to further refine our technology by optimizing the loading density of stem cells and by incorporating the topographical guidance structures. Given the focus i.e., early nerve regeneration of the present study, we measured only early axonal regeneration. However, the data and knowledge obtained in the present study create strong rationale for the further evaluation of functional restoration, i.e., behavioral recovery, electrophysiological recovery and re-innervation in a long-term study.

The delivery of cells within a conduit is a subject of research and the most optimal delivery system remains to be determined. Particularly for FNC, only a few methods have been evaluated separately and no systematic comparison was reported so far. In general, fibrin solution loaded with cells can be injected into the lumen of the FNC [27]. An alternative approach is to integrate the fibrin solution loaded with the cells into the main structure of the FNC's wall by polymerization [26]. The least interesting way is to suspend the cells in a carrier medium and to place them into the lumen of the conduit [30], which usually results in leakage of cells.

In our study, we tested the first two approaches by delivering the NGF-stimulated ASC either by FNC assisted intramural or by intraluminal loading. For this, fibrin conduits measuring 14 mm in length, 1 mm in wall thickness and 2 mm in lumen were produced for bridging a 10-mm nerve gap injury. It is widely accepted that fibrin hydrogels possess macro porous structures with a pore size of 10–20 μm [59]. Therefore, conduits used in our study may possess pore sizes in the range of 2–4 μm and 10–20 μm respectively for intramural and intraluminal fibrin structures that can naturally facilitate the diffusion of ASC secretome. ASC based exosomes and microvesicles are in the range of a sub-micrometer size (i.e., 150 nm to 200 nm) [49,60]. Thus, the macro porous fibrin hydrogels enable easy passage of nanometer-sized vesicles and related soluble factors. However, the density of transplanted cells within the hydrogel microenvironment may account for different outcomes achieved through different delivery routes in the present study. In the case of the intramural delivery system, 2 million cells were loaded in the 600 μL of carrier fibrin hydrogel in contrast to an intraluminal delivery system where 2 million cells were loaded in 40 μL of carrier hydrogel. Although homogeneity of cell treatment (i.e., the dose of the total cells) for animals was ensured, the high density of cells within the lumen of the fibrin nerve conduit may result in the high concentrations of the local growth factors. Thus, the highly enriched luminal microenvironment of the intraluminal delivery system may impede the speed of axonal regeneration in contrast to the intramural delivery system. Taken together, the underlying structural and cellular factors may explain appropriately the better outcome achieved by the intramural cell delivery system in the present study.

Fibrinogen concentrations used in the present study for both intramural and intraluminal constructs were based on our earlier research work and no-detrimental effects were found when the fibrin matrix was used alone. As we previously reported [29], ASC showed survival signals up to 14 days after transplantation through nerve conduit. These findings indicate that the regenerative effects of transplanted ASC are mediated by the initial boost of released growth factors. The cell delivery route dependent outcome observed in our present study may partly be attributed to the potency of the transplanted cells. Further investigation on the spatiotemporal profile of the transplanted cells may provide important understanding and knowledge for the optimization of ASC therapies. The present study was conducted using rat-derived ASC, therefore further studies are required using human ASC, in order to translate these findings into clinical settings.5. Conclusions

We report on a growth factor based strategy for ex-vivo stimulation of ASC and show the evidence for the enhanced neurotrophic potency of ASC in response to NGF-stimulation, but not VEGF, to promote axonal regeneration in vitro and in vivo. Furthermore, our study reveals the importance of a fibrin-hydrogel conduit assisted intramural delivery system in improving the therapeutic efficacy of ASC for nerve regeneration. Together, these findings provide new knowledge and important insights for the development of ASC based therapies for treating nerve gap-injuries. Further studies are required for assessing the long-term impact of the ex-vivo stimulated ASC on anatomical and functional nerve regeneration, and for understanding the secretome profile of the ASC resulting from various stimuli-conditions.

Supplementary Materials: The following are available online at http://www.mdpi.com/2306-5354/7/2/42/s1, Figure S1: Stem cell characterization, Figure S2: (A) Immunostaining of Schwann cell growth. (B) Quantitative analysis of Schwann cell growth.

Author Contributions: S.M. contributed conception and design of the study; S.M. supervised the project. K.M.P. and L.D. conducted both in vitro and in vivo experiments; S.M., D.F.K., R.G., D.J.S. and K.M.P. contributed for important intellectual interpretation of the study; S.M. and K.M.P. performed the statistical analysis; K.M.P. wrote the first draft of the manuscript; S.M. revised the manuscript. All the authors contributed to manuscript revision, read and approved the manuscript submission. All authors have read and agreed to the published version of the manuscript.

Funding: We thank the enabling research grant support by EUROSTAR to Madduri Srinivas and by surgery foundation, University Hospital Basel to Madduri Srinivas, Daniel Kalbermatten and Raphael Guzman.

Acknowledgments: We thank Michael Abanto and Loic Sauteur of the DBM Microscopy core Facility, University of Basel for the excellent technical support.

Conflicts of Interest: Authors declare no conflict of interest.

References

1. Lundborg, G. A 25-year perspective of peripheral nerve surgery: Evolving neuroscientific concepts and clinical significance. *J. Hand Surg. Am.* **2000**, *25*, 391–414. [CrossRef] [PubMed]
2. Wiberg, M.; Terenghi, G. Will it be possible to produce peripheral nerves? *Surg. Technol. Int.* **2003**, *11*, 303–310. [PubMed]

3. Webber, C.; Zochodne, D. The nerve regenerative microenvironment: Early behavior and partnership of axons and Schwann cells. *Exp. Neurol.* **2010**, *223*, 51–59. [CrossRef] [PubMed]
4. Allodi, I.; Udina, E.; Navarro, X. Specificity of peripheral nerve regeneration: Interactions at the axon level. *Prog. Neurobiol.* **2012**, *98*, 16–37. [CrossRef] [PubMed]
5. Widgerow, A.D.; Salibian, A.A.; Lalezari, S.; Evans, G.R. Neuromodulatory nerve regeneration: Adipose tissue-derived stem cells and neurotrophic mediation in peripheral nerve regeneration. *J. Neurosci. Res.* **2013**, *91*, 1517–1524. [CrossRef] [PubMed]
6. Nakagami, H.; Maeda, K.; Morishita, R.; Iguchi, S.; Nishikawa, T.; Takami, Y.; Kikuchi, Y.; Saito, Y.; Tamai, K.; Ogihara, T.; et al. Novel autologous cell therapy in ischemic limb disease through growth factor secretion by cultured adipose tissue-derived stromal cells. *Arterioscler. Thromb. Vasc. Biol.* **2005**, *25*, 2542–2547. [CrossRef]
7. di Summa, P.G.; Kalbermatten, D.F.; Pralong, E.; Raffoul, W.; Kingham, P.J.; Terenghi, G. Long-term in vivo regeneration of peripheral nerves through bioengineered nerve grafts. *Neuroscience* **2011**, *181*, 278–291. [CrossRef]
8. Dezawa, M.; Takahashi, I.; Esaki, M.; Takano, M.; Sawada, H. Sciatic nerve regeneration in rats induced by transplantation of in vitro differentiated bone-marrow stromal cells. *Eur. J. Neurosci.* **2001**, *14*, 1771–1776. [CrossRef]
9. Tohill, M.; Mantovani, C.; Wiberg, M.; Terenghi, G. Rat bone marrow mesenchymal stem cells express glial markers and stimulate nerve regeneration. *Neurosci. Lett.* **2004**, *362*, 200–203. [CrossRef]
10. Keilhoff, G.; Goihl, A.; Langnase, K.; Fansa, H.; Wolf, G. Transdifferentiation of mesenchymal stem cells into Schwann cell-like myelinating cells. *Eur. J. Cell Biol.* **2006**, *85*, 11–24. [CrossRef]
11. Caddick, J.; Kingham, P.J.; Gardiner, N.J.; Wiberg, M.; Terenghi, G. Phenotypic and functional characteristics of mesenchymal stem cells differentiated along a Schwann cell lineage. *Glia* **2006**, *54*, 840–849. [CrossRef] [PubMed]
12. De Ugarte, D.A.; Morizono, K.; Elbarbary, A.; Alfonso, Z.; Zuk, P.A.; Zhu, M.; Dragoo, J.L.; Ashjian, P.; Thomas, B.; Benhaim, P.; et al. Comparison of multi-lineage cells from human adipose tissue and bone marrow. *Cells Tissues Organs* **2003**, *174*, 101–109. [CrossRef] [PubMed]
13. Strem, B.M.; Hicok, K.C.; Zhu, M.; Wulur, I.; Alfonso, Z.; Schreiber, R.E.; Fraser, J.K.; Hedrick, M.H. Multipotential differentiation of adipose tissue-derived stem cells. *Keio J. Med.* **2005**, *54*, 132–141. [CrossRef]
14. Cruciani, S.; Santaniello, S.; Montella, A.; Ventura, C.; Maioli, M. Orchestrating stem cell fate: Novel tools for regenerative medicine. *World J. Stem Cells* **2019**, *11*, 464–475. [CrossRef] [PubMed]
15. Mazzini, L.; Ferrari, D.; Andjus, P.R.; Buzanska, L.; Cantello, R.; De Marchi, F.; Gelati, M.; Giniatullin, R.; Glover, J.C.; Grilli, M.; et al. Advances in stem cell therapy for amyotrophic lateral sclerosis. *Expert Opin. Biol. Ther.* **2018**, *18*, 865–881. [CrossRef]
16. Kern, S.; Eichler, H.; Stoeve, J.; Kluter, H.; Bieback, K. Comparative analysis of mesenchymal stem cells from bone marrow, umbilical cord blood, or adipose tissue. *Stem Cells (Dayt. Ohio)* **2006**, *24*, 1294–1301. [CrossRef]
17. Zuk, P.A.; Zhu, M.; Ashjian, P.; De Ugarte, D.A.; Huang, J.I.; Mizuno, H.; Alfonso, Z.C.; Fraser, J.K.; Benhaim, P.; Hedrick, M.H. Human adipose tissue is a source of multipotent stem cells. *Mol. Biol. Cell* **2002**, *13*, 4279–4295. [CrossRef]
18. El Atat, O.; Antonios, D.; Hilal, G.; Hokayem, N.; Abou-Ghoch, J.; Hashim, H.; Serhal, R.; Hebbo, C.; Moussa, M.; Alaaeddine, N. An Evaluation of the Stemness, Paracrine, and Tumorigenic Characteristics of Highly Expanded, Minimally Passaged Adipose-Derived Stem Cells. *PLoS ONE* **2016**, *11*, e0162332. [CrossRef]
19. Aust, L.; Devlin, B.; Foster, S.J.; Halvorsen, Y.D.; Hicok, K.; du Laney, T.; Sen, A.; Willingmyre, G.D.; Gimble, J.M. Yield of human adipose-derived adult stem cells from liposuction aspirates. *Cytotherapy* **2004**, *6*, 7–14. [CrossRef]
20. McIntosh, K.R.; Lopez, M.J.; Borneman, J.N.; Spencer, N.D.; Anderson, P.A.; Gimble, J.M. Immunogenicity of allogeneic adipose-derived stem cells in a rat spinal fusion model. *Tissue Eng. Part A* **2009**, *15*, 2677–2686. [CrossRef]
21. Kingham, P.J.; Kalbermatten, D.F.; Mahay, D.; Armstrong, S.J.; Wiberg, M.; Terenghi, G. Adipose-derived stem cells differentiate into a Schwann cell phenotype and promote neurite outgrowth in vitro. *Exp. Neurol.* **2007**, *207*, 267–274. [CrossRef] [PubMed]

22. Kingham, P.J.; Kolar, M.K.; Novikova, L.N.; Novikov, L.N.; Wiberg, M. Stimulating the neurotrophic and angiogenic properties of human adipose-derived stem cells enhances nerve repair. *Stem Cells Dev.* **2014**, *23*, 741–754. [CrossRef] [PubMed]
23. Bucan, V.; Vaslaitis, D.; Peck, C.T.; Strauss, S.; Vogt, P.M.; Radtke, C. Effect of Exosomes from Rat Adipose-Derived Mesenchymal Stem Cells on Neurite Outgrowth and Sciatic Nerve Regeneration After Crush Injury. *Mol. Neurobiol.* **2019**, *56*, 1812–1824. [CrossRef] [PubMed]
24. Georgiou, M.; Golding, J.P.; Loughlin, A.J.; Kingham, P.J.; Phillips, J.B. Engineered neural tissue with aligned, differentiated adipose-derived stem cells promotes peripheral nerve regeneration across a critical sized defect in rat sciatic nerve. *Biomaterials* **2015**, *37*, 242–251. [CrossRef] [PubMed]
25. Sowa, Y.; Kishida, T.; Imura, T.; Numajiri, T.; Nishino, K.; Tabata, Y.; Mazda, O. Adipose-Derived Stem Cells Promote Peripheral Nerve Regeneration In Vivo without Differentiation into Schwann-Like Lineage. *Plast. Reconstr. Surg.* **2016**, *137*, 318e–330e. [CrossRef]
26. Saller, M.M.; Huettl, R.E.; Mayer, J.M.; Feuchtinger, A.; Krug, C.; Holzbach, T.; Volkmer, E. Validation of a novel animal model for sciatic nerve repair with an adipose-derived stem cell loaded fibrin conduit. *Neural Regen. Res.* **2018**, *13*, 854–861. [CrossRef]
27. Kappos, E.A.; Engels, P.E.; Tremp, M.; Meyer zu Schwabedissen, M.; di Summa, P.; Fischmann, A.; von Felten, S.; Scherberich, A.; Schaefer, D.J.; Kalbermatten, D.F. Peripheral Nerve Repair: Multimodal Comparison of the Long-Term Regenerative Potential of Adipose Tissue-Derived Cells in a Biodegradable Conduit. *Stem Cells Dev.* **2015**, *24*, 2127–2141. [CrossRef]
28. Klein, S.M.; Vykoukal, J.; Li, D.P.; Pan, H.L.; Zeitler, K.; Alt, E.; Geis, S.; Felthaus, O.; Prantl, L. Peripheral Motor and Sensory Nerve Conduction following Transplantation of Undifferentiated Autologous Adipose Tissue-Derived Stem Cells in a Biodegradable U.S. Food and Drug Administration-Approved Nerve Conduit. *Plast. Reconstr. Surg.* **2016**, *138*, 132–139. [CrossRef]
29. Erba, P.; Mantovani, C.; Kalbermatten, D.F.; Pierer, G.; Terenghi, G.; Kingham, P.J. Regeneration potential and survival of transplanted undifferentiated adipose tissue-derived stem cells in peripheral nerve conduits. *J. Plast. Reconstr. Aesthetic Surg. Jpras* **2010**, *63*, e811–e817. [CrossRef]
30. di Summa, P.G.; Kingham, P.J.; Raffoul, W.; Wiberg, M.; Terenghi, G.; Kalbermatten, D.F. Adipose-derived stem cells enhance peripheral nerve regeneration. *J. Plast. Reconstr. Aesthetic Surg. Jpras* **2010**, *63*, 1544–1552. [CrossRef]
31. Hoyng, S.A.; De Winter, F.; Gnavi, S.; de Boer, R.; Boon, L.I.; Korvers, L.M.; Tannemaat, M.R.; Malessy, M.J.; Verhaagen, J. A comparative morphological, electrophysiological and functional analysis of axon regeneration through peripheral nerve autografts genetically modified to overexpress BDNF, CNTF, GDNF, NGF, NT3 or VEGF. *Exp. Neurol.* **2014**, *261*, 578–593. [CrossRef] [PubMed]
32. Faroni, A.; Mobasseri, S.A.; Kingham, P.J.; Reid, A.J. Peripheral nerve regeneration: Experimental strategies and future perspectives. *Adv. Drug Deliv. Rev.* **2015**, *82–83*, 160–167. [CrossRef] [PubMed]
33. Belanger, K.; Dinis, T.M.; Taourirt, S.; Vidal, G.; Kaplan, D.L.; Egles, C. Recent Strategies in Tissue Engineering for Guided Peripheral Nerve Regeneration. *Macromol. Biosci.* **2016**, *16*, 472–481. [CrossRef] [PubMed]
34. Madduri, S.; Papaloizos, M.; Gander, B. Synergistic effect of GDNF and NGF on axonal branching and elongation in vitro. *Neurosci. Res.* **2009**, *65*, 88–97. [CrossRef] [PubMed]
35. Hobson, M.I. Increased vascularisation enhances axonal regeneration within an acellular nerve conduit. *Ann. R. Coll. Surg. Engl.* **2002**, *84*, 47–53. [PubMed]
36. Sondell, M.; Sundler, F.; Kanje, M. Vascular endothelial growth factor is a neurotrophic factor which stimulates axonal outgrowth through the flk-1 receptor. *Eur. J. Neurosci.* **2000**, *12*, 4243–4254. [CrossRef]
37. Sanchez, A.; Wadhwani, S.; Grammas, P. Multiple neurotrophic effects of VEGF on cultured neurons. *Neuropeptides* **2010**, *44*, 323–331. [CrossRef]
38. Calvo, P.M.; Pastor, A.M.; de la Cruz, R.R. Vascular endothelial growth factor: An essential neurotrophic factor for motoneurons? *Neural Regen. Res.* **2018**, *13*, 1181–1182. [CrossRef]
39. Muratori, L.; Gnavi, S.; Fregnan, F.; Mancardi, A.; Raimondo, S.; Perroteau, I.; Geuna, S. Evaluation of Vascular Endothelial Growth Factor (VEGF) and Its Family Member Expression After Peripheral Nerve Regeneration and Denervation. *Anat. Rec.* **2018**, *301*, 1646–1656. [CrossRef]
40. Madduri, S.; di Summa, P.; Papaloizos, M.; Kalbermatten, D.; Gander, B. Effect of controlled co-delivery of synergistic neurotrophic factors on early nerve regeneration in rats. *Biomaterials* **2010**, *31*, 8402–8409. [CrossRef]

41. Wood, M.D.; MacEwan, M.R.; French, A.R.; Moore, A.M.; Hunter, D.A.; Mackinnon, S.E.; Moran, D.W.; Borschel, G.H.; Sakiyama-Elbert, S.E. Fibrin matrices with affinity-based delivery systems and neurotrophic factors promote functional nerve regeneration. *Biotechnol. Bioeng.* **2010**, *106*, 970–979. [CrossRef] [PubMed]
42. Kalbermatten, D.F.; Pettersson, J.; Kingham, P.J.; Pierer, G.; Wiberg, M.; Terenghi, G. New fibrin conduit for peripheral nerve repair. *J. Reconstr. Microsurg.* **2009**, *25*, 27–33. [CrossRef] [PubMed]
43. Pettersson, J.; Kalbermatten, D.; McGrath, A.; Novikova, L.N. Biodegradable fibrin conduit promotes long-term regeneration after peripheral nerve injury in adult rats. *J. Plast. Reconstr. Aesthetic Surg. Jpras* **2010**, *63*, 1893–1899. [CrossRef] [PubMed]
44. Zhang, R.; Rosen, J.M. The role of undifferentiated adipose-derived stem cells in peripheral nerve repair. *Neural Regen. Res.* **2018**, *13*, 757–763. [CrossRef]
45. Orbay, H.; Uysal, A.C.; Hyakusoku, H.; Mizuno, H. Differentiated and undifferentiated adipose-derived stem cells improve function in rats with peripheral nerve gaps. *J. Plast. Reconstr. Aesthetic Surg. Jpras* **2012**, *65*, 657–664. [CrossRef]
46. Watanabe, Y.; Sasaki, R.; Matsumine, H.; Yamato, M.; Okano, T. Undifferentiated and differentiated adipose-derived stem cells improve nerve regeneration in a rat model of facial nerve defect. *J. Tissue Eng. Regen. Med.* **2017**, *11*, 362–374. [CrossRef]
47. Radtke, C.; Schmitz, B.; Spies, M.; Kocsis, J.D.; Vogt, P.M. Peripheral glial cell differentiation from neurospheres derived from adipose mesenchymal stem cells. *Int. J. Dev. Neurosci. Off. J. Int. Soc. Dev. Neurosci.* **2009**, *27*, 817–823. [CrossRef]
48. Haghighat, A.; Akhavan, A.; Hashemi-Beni, B.; Deihimi, P.; Yadegari, A.; Heidari, F. Adipose derived stem cells for treatment of mandibular bone defects: An autologous study in dogs. *Dent. Res. J.* **2011**, *8*, S51–S57.
49. Fang, Y.; Zhang, Y.; Zhou, J.; Cao, K. Adipose-derived mesenchymal stem cell exosomes: A novel pathway for tissues repair. *Cell Tissue Bank.* **2019**, *20*, 153–161. [CrossRef]
50. Santiago, L.Y.; Clavijo-Alvarez, J.; Brayfield, C.; Rubin, J.P.; Marra, K.G. Delivery of adipose-derived precursor cells for peripheral nerve repair. *Cell Transplant.* **2009**, *18*, 145–158. [CrossRef]
51. Salgado, A.J.; Reis, R.L.; Sousa, N.J.; Gimble, J.M. Adipose tissue derived stem cells secretome: Soluble factors and their roles in regenerative medicine. *Curr. Stem Cell Res. Ther.* **2010**, *5*, 103–110. [CrossRef] [PubMed]
52. Marconi, S.; Castiglione, G.; Turano, E.; Bissolotti, G.; Angiari, S.; Farinazzo, A.; Constantin, G.; Bedogni, G.; Bedogni, A.; Bonetti, B. Human adipose-derived mesenchymal stem cells systemically injected promote peripheral nerve regeneration in the mouse model of sciatic crush. *Tissue Eng. Part A* **2012**, *18*, 1264–1272. [CrossRef] [PubMed]
53. Bellei, B.; Migliano, E.; Tedesco, M.; Caputo, S.; Papaccio, F.; Lopez, G.; Picardo, M. Adipose tissue-derived extracellular fraction characterization: Biological and clinical considerations in regenerative medicine. *Stem Cell Res. Ther.* **2018**, *9*, 207. [CrossRef] [PubMed]
54. Vasilev, G.; Ivanova, M.; Ivanova-Todorova, E.; Tumangelova-Yuzeir, K.; Krasimirova, E.; Stoilov, R.; Kyurkchiev, D. Secretory factors produced by adipose mesenchymal stem cells downregulate Th17 and increase Treg cells in peripheral blood mononuclear cells from rheumatoid arthritis patients. *Rheumatol. Int.* **2019**, *39*, 819–826. [CrossRef]
55. Madduri, S.; Papaloizos, M.; Gander, B. Trophically and topographically functionalized silk fibroin nerve conduits for guided peripheral nerve regeneration. *Biomaterials* **2010**, *31*, 2323–2334. [CrossRef]
56. Lee, J.Y.; Giusti, G.; Friedrich, P.F.; Bishop, A.T.; Shin, A.Y. Effect of Vascular Endothelial Growth Factor Administration on Nerve Regeneration after Autologous Nerve Grafting. *J. Reconstr. Microsurg.* **2016**, *32*, 183–188. [CrossRef]
57. Pereira Lopes, F.R.; Lisboa, B.C.; Frattini, F.; Almeida, F.M.; Tomaz, M.A.; Matsumoto, P.K.; Langone, F.; Lora, S.; Melo, P.A.; Borojevic, R.; et al. Enhancement of sciatic nerve regeneration after vascular endothelial growth factor (VEGF) gene therapy. *Neuropathol. Appl. Neurobiol.* **2011**, *37*, 600–612. [CrossRef]
58. Xia, B.; Lv, Y. Dual-delivery of VEGF and NGF by emulsion electrospun nanofibrous scaffold for peripheral nerve regeneration. *Mater. Sci. Eng. CMater. Biol. Appl.* **2018**, *82*, 253–264. [CrossRef]

59. Lisi, A.; Briganti, E.; Ledda, M.; Losi, P.; Grimaldi, S.; Marchese, R.; Soldani, G. A combined synthetic-fibrin scaffold supports growth and cardiomyogenic commitment of human placental derived stem cells. *PLoS ONE* **2012**, *7*, e34284. [CrossRef]
60. Mohammadi, M.R.; Riazifar, M.; Pone, E.J.; Yeri, A.; Van Keuren-Jensen, K.; Lasser, C.; Lotvall, J.; Zhao, W. Isolation and characterization of microvesicles from mesenchymal stem cells. *Methods (San DiegoCalif.)* **2019**. [CrossRef]

Article

Impact of Dual Cell Co-culture and Cell-conditioned Media on Yield and Function of a Human Olfactory Cell Line for Regenerative Medicine

Rachael Wood [1,2], Pelin Durali [1] and Ivan Wall [1,2,3,*]

1 Department of Biochemical Engineering, University College London, Torrington Place, London WC1E 7JE, UK; rachael.wood.14@ucl.ac.uk (R.W.); pelin.durali.14@alumni.ucl.ac.uk (P.D.)
2 School of Life & Health Sciences, Aston University, Aston Triangle, Birmingham B4 7ET, UK
3 Institute of Tissue Regeneration Engineering (ITREN), Dankook University, Cheonan 31116, Korea
* Correspondence: i.wall@aston.ac.uk

Received: 4 March 2020; Accepted: 10 April 2020; Published: 12 April 2020

Abstract: Olfactory ensheathing cells (OECs) are a promising candidate therapy for neuronal tissue repair. However, appropriate priming conditions to drive a regenerative phenotype are yet to be determined. We first assessed the effect of using a human fibroblast feeder layer and fibroblast conditioned media on primary rat olfactory mucosal cells (OMCs). We found that OMCs cultured on fibroblast feeders had greater expression of the key OEC marker p75NTR (25.1 ± 10.7 cells/mm^2) compared with OMCs cultured on laminin (4.0 ± 0.8 cells/mm^2, $p = 0.001$). However, the addition of fibroblast-conditioned media (CM) resulted in a significant increase in Thy1.1 (45.9 ± 9.0 cells/mm^2 versus 12.5 ± 2.5 cells/mm^2 on laminin, $p = 0.006$), an undesirable cell marker as it is regarded to be a marker of contaminating fibroblasts. A direct comparison between human feeders and GMP cell line Ms3T3 was then undertaken. Ms3T3 cells supported similar p75NTR levels (10.7 ± 5.3 cells/mm^2) with significantly reduced Thy1.1 expression (4.8 ± 2.1 cells/mm^2). Ms3T3 cells were used as feeder layers for human OECs to determine whether observations made in the rat model were conserved. Examination of the OEC phenotype (S100β expression and neurite outgrowth from NG108-15 cells) revealed that co-culture with fibroblast feeders had a negative effect on human OECs, contrary to observations of rat OECs. CM negatively affected rat and human OECs equally. When the best and worst conditions in terms of supporting S100β expression were used in NG108-15 neuron co-cultures, those with the highest S100β expression resulted in longer and more numerous neurites (22.8 ± 2.4 μm neurite length/neuron for laminin) compared with the lowest S100β expression (17.9 ± 1.1 μm for Ms3T3 feeders with CM). In conclusion, this work revealed that neither dual co-culture nor fibroblast-conditioned media support the regenerative OEC phenotype. In our case, a preliminary rat model was not predictive of human cell responses.

Keywords: olfactory ensheathing cells; spinal cord injury; neural regeneration; cell therapies

---

1. Introduction

Spinal cord injury (SCI) is a devastating injury to the central nervous system (CNS) that affects 250,000–500,000 new people worldwide every year. People with an SCI are 2–5 times more likely to die prematurely [1], and lifetime costs are estimated between US$1 and US$5 million, excluding indirect costs such as a loss in wages [2]. The majority of patients will suffer partial or complete paralysis [3] and can also suffer from chronic neuropathic pain syndromes, which have a serious effect on the quality of life [4]. Following SCI, inflammation and cell death ensue, and eventually, scar tissue forms, which is the main impediment to spontaneous regeneration, as the axons that do sprout and re-grow are unable to path-find through the scar to reach their target [5–9].

Unlike the rest of the body, the CNS does not regenerate. An exception is the olfactory system, which retains its ability to regenerate throughout adult life, due to the presence of a special type of glial cell, the olfactory ensheathing cell (OEC) [10]. These cells have been studied for potential use in spinal cord repair [11] due to their natural role in regenerating and guiding olfactory receptor neurons from the peripheral nervous system (PNS) into the CNS. Mucosal OECs can be biopsied via a minimally invasive intranasal approach, and although mucosa OECs tend to have lower yields and purity, they are more clinically attractive than their counterparts, bulb OECs, which are found in the lining of the brain [12].

Characterisation of OECs is commonly based on the expression of the neurotrophic receptor, p75NTR, in cell populations isolated from the olfactory mucosa or bulb [11,13–15]. p75NTR is a receptor that induces neurite outgrowth and cellular survival [16]; however, it is not a definitive marker for OECs [17]. In addition to p75NTR, glial cell marker S100β is commonly used as a positive marker for OECs [18], and Thy1.1 and fibronectin (Fn) are used as markers for contaminating fibroblasts [11,19–21].

For an OEC therapy to be developed, more understanding is needed of how to control the growth of functional subsets that express glial markers p75NTR and S100β and also how accurately these putative identity markers are predictive of function. If we better understand the markers that predict regenerative potential, we can then make strides toward enriching cells expressing those markers.

Due to the rapid onset of damage that occurs during an SCI, surgical options have to be enacted within 24 h of the injury [22], and by three weeks, the majority of the damage has occurred, so it is important to administer any cell therapy before this point [23]. A key issue with OECs is that their yield and purity variance cannot be predicted. Due to this, allogeneic cell therapy is seen as the most promising option as an off-the-shelf therapy that can be administered in the vital period before severe secondary damage sets in [24,25]. With the variability between patients, there would be no guarantee that enough functional cells could be produced in the required time frame, let alone all the release testing carried out to ensure the patient's safety. Previously, our group generated conditionally immortalised human mucosa-derived OEC cell lines [26] toward the goal of producing an off-the-shelf therapy for SCI. In this study, we aim to compare our human cell line against primary rat cells to determine the best conditions for OEC culture and identify the relevance of animal studies in this area.

2. Materials and Methods

2.1. Primary Cell Isolation and Culture

Mucosae were isolated following the optimum protocol previously identified [27]. The mucosae were dissected from three adult female Sprague-Dawley 200–250g rats, which were euthanized by carbon dioxide asphyxiation (Schedule 1 method [28]) according to the U.K. Animals (Scientific Procedures) Act 1986. Each rat provided two mucosae. Each mucosa was placed in DMEM/F12 media to be transported to the laboratory. The mucosae were placed in Hanks' Balanced Salt Solution (HBSS, Gibco Life Technologies, Gaithersburg, MD, USA) with 1% P/S (10,000 units penicillin, 10 mg streptomycin/mL) in a petri dish and washed by gently wiping each mucosa across a spatula. After washing, the mucosae were placed on a petri dish and cut up into small pieces, then placed in a 5 mL dispase II (2.4 units/mL, Sigma-Aldrich, UK) solution for 45 min at 37 °C in order to digest the tissue enzymatically.

The mucosa and dispase solution was placed in the centrifuge for five minutes at 400× *g*. The supernatant was discarded, and the cells and tissue were re-suspended in 5 mL collagenase (Type I solution, 0.05%, Sigma-Aldrich, UK) for 15 min at 37 °C. Every five minutes, the collagenase was taken out of the incubator and mechanically triturated briefly before being placed back in the incubator. At the 15 minute point, the collagenase suspension was placed in the centrifuge for five minutes at 400× *g*. The cells were re-suspended in 7 mL DMEM/F12 media (2% FBS, 1% P/S) in a T25 flask at 37 °C in 5% CO_2 for 24 h. All cell cultures were carried out at 37 °C and 5% CO_2.

The cells were placed in a tissue culture flask for 24 h as a differential adhesion step. The purpose of this step was to decrease the amount of contaminating fibroblasts in the culture. Fibroblasts have a faster adhesion time than OECs. After 24 h, most fibroblasts will adhere to the tissue culture plastic, whereas OECs will still be in suspension [27]. After 24 h, the suspension was replated onto laminin-coated (20 µg/mL, Sigma-Aldrich, UK) wells. The wells were coated for four hours at 37 °C. The laminin was then removed, and the cells were plated while the matrix was still wet.

2.2. Cell Line Generation and Culture

The feeder layer (HuG418) was prepared from cells isolated from the human olfactory mucosa. The cells used were thawed from a pre-existing cell line prepared by Dr. Melanie Georgiou (UCL). The protocol followed by Dr. Melanie Georgiou was summarised in the paper by Pollock et al. [29]. The human mucosa OEC line was generated using the same technology; however, the purification steps were carried out before immortalisation to ensure a purer OEC population. The generation and culture conditions of this line (PA5) were covered by Santiago-Toledo et al. [26], and the culture conditions remained the same for HuG418.

2.3. Feeder Layer Generation

Feeder layers were prepared by removing the media from the T75 flask and replacing with 4 mL mitomycin C (MMC) (Sigma-Aldrich, UK) for 2 h at 37 °C to inactivate the cells. After two hours, the MMC was removed from the cells, and the cells were washed with PBS (Lonza). The cells were removed from the flask using Trypsin/EDTA (Sigma) and placed in the centrifuge for five minutes at 400× *g*. After discarding the supernatant, cells were re-suspended in DMEM/F12 and plated up at 12×10^3 cells/cm^2. This density was used as it gave a good coverage of the well whilst allowing space for the OECs on the well surface. Feeders remained in the incubator for two days before any cells were plated on them.

2.4. Collection of Conditioned Media

Conditioned media (CM) was collected from HuG418 cells after two days of culture. The media was centrifuged for five minutes at 400× *g*. CM was mixed at 1:1 with fresh media and added to the required conditions.

2.5. Culture of PA5 Cells

For the experiments involving PA5 cells, wells were coated either with Ms3T3 feeders, poly-l-lysine (PLL), or laminin. PA5 cells were plated at 6000 cells/cm^2 and cultured for five days. After five days, cells were fixed and stained for S100β and fibronectin. Neurotrophic factor 3 (NT-3) was added to some media conditions at 50 ng/mL as the literature indicated that NT-3 encourages the neurological repair phenotype in OECs [27].

2.6. In Vitro Co-culture with NG108-15 Neurons

NG108-15 neurons were either grown alone on Ms3T3 feeders (as a control) or on a layer of PA5 cells that were seeded onto either laminin or Ms3T3 feeders. NG108-15 neurons were seeded at 500 cells/cm^2 onto the laminin-coated plates or feeder layers. Co-culture was carried out for five days with a media change occurring on Day 3 following the protocol laid out by Santiago-Toledo et al. [26]. Cells were fixed with 4% PFA at room temperature (RT), and cells were stained to detect β-III tubulin.

2.7. Immunocytochemistry (ICC)

Cells were fixed with 4% paraformaldehyde (PFA, Sigma-Aldrich, UK) for 20 min at RT. The PFA was removed, and the cells were washed three times with PBS. Three washes with PBS were carried out in between every step of the immuno-staining in order to remove and dilute any traces of the previous reagent. One wash was defined as filling the well with PBS and leaving for five minutes. Zero-point-two-five percent Triton-X (Sigma-Aldrich, UK) was added to each well and left for

20 min at RT. All solutions were made up in PBS. Five percent goat serum solution (Dako) was used as the blocking solution for 30 min at RT. Primary antibodies were prepared at a dilution of 1:200 and added for 90 min at RT. The primary antibodies used were as follows: rabbit anti-p75NTR (Millipore), mouse anti-Thy1.1 (Millipore), rabbit anti-S100β (Dako), mouse anti-fibronectin (Sigma-Aldrich, UK), and mouse anti-βIII tubulin (Sigma-Aldrich, UK). The negative control was subjected to the same treatment as other wells, the exception being the exclusion of the primary antibody. The secondary antibodies (DyLight® 488 or 594, goat anti-rabbit IgG (H + L) and goat anti-mouse IgG (H + L), Vector Laboratories, Burlingame, CA, USA) were prepared at 1:200 and Hoechst (Sigma-Aldrich, UK) at 1:1000 and incubated for 45 min at RT. Fluorescent imaging was carried out on an EVOS FL microscope (Life Technologies).

Five images were taken per well and were taken in a cross formation around the centre of the well. Two to three wells were stained per condition. After imaging, each channel was examined to assess whether the images taken were representative of the well. Brightness and shutter speed were set by identifying a cell that was determined to be positive and ensuring the background was dark to prevent overexposure of the cells. These settings were held constant between images for an experiment to ensure the images could be directly compared.

Images were analysed using ImageJ. A colour threshold was set to identify a positive cell, and only cells that were identified to be above this threshold were counted as positive. Positive cells were counted, and the results were calculated as the proportion of cells positive for the marker and the yield of positive cells over the imaged area.

2.8. Circularity Analysis

A macro was written in ImageJ in order to analyse the morphology of the cells. Circularity was used as the defining value. The macro was written in JavaScript to automate a calculation method that already exists in ImageJ. The "adjust threshold" window opens automatically when the macro is run and can be adjusted to select the area of interest. Any cell that was larger than a set threshold (this value can be adjusted depending on whether the nucleus or the whole cell is being analysed) was discarded to prevent cell clusters being included in the calculation. In addition to this, any cells on the boundary of the image were discarded so they did not contribute to false readings. The circularity was calculated as in Equation (1), where A is the area (pixels^2) and P is the perimeter (pixels).

$$Circularity = \frac{4A\pi}{p^2} \tag{1}$$

A circularity of 1 indicates a perfect circle, and a circularity of 0 indicates a straight line. When the macro was run, a printout of the circularities was displayed. These data were exported to GraphPad® and plotted as a histogram.

2.9. Neuronal Growth Analysis

According to the literature, five days of co-culture are optimal for observing enhanced neurite outgrowth compared with three days of co-culture [27]. After five days of co-culture, cells were labelled with antibodies against βIII-tubulin. Neuron number, neurite number, and neurite length were quantified using the NeuronJ plugin in ImageJ [30]. Neurites were traced and measured within the software and exported to GraphPad® for analysis.

2.10. Statistical Analysis and Data Accessibility

Data are presented as the mean ± standard error of the mean (SEM). For the majority of the data, one-way ANOVA was used to determine statistically significant differences, and the Bonferroni correction was carried out using GraphPad® software to calculate p-values. When there were only two sets of data to compare, the Shapiro–Wilk test was used to test for normality. When data were normally distributed, the Tukey test was used to test for significance. When data were not normally distributed, the Kruskal–Wallis test was used. The Bonferroni correction was not used for sets of data that only compared two conditions as there was no need to correct for multiple comparisons. For the comparison of histograms, the Kolmogorov–Smirnov test was carried out.

In this work, one experimental repeat (n) was defined as cells taken from different flasks/groups of animals.

Statistics are only reported in the text when there are significant differences at $\alpha = 0.05$. In all figures, a single asterisk (*) indicates $p < 0.05$, two asterisks (**) $p < 0.01$, and three asterisks (***) $p < 0.001$.

The datasets generated and analysed during this study are available from the corresponding author upon reasonable request.

3. Results and Discussion

3.1. Human Feeders Encourage an Increase in p75NTR and Spindle-Shaped Cells in Rat OECs

The ICC was quantified using yield (positive cells per mm^2), but not purity, as the presence of the feeders in only some conditions would make any assessment of purity misleading. It can be observed from Figure 1B,E,H,K,M that the addition of CM significantly increased the expression of Thy1.1. When CM was added to OECs cultured on feeders, an increase in Thy1.1 over and above the higher expression induced by feeders with standard media was observed (25.7 ± 12.4 cells/mm^2 on feeders with CM from 14.5 ± 4.8 cells/mm^2 on feeders with standard media). The increase in Thy1.1 expression observed on laminin with CM (45.9 ± 9.0 cells/mm^2) indicated that HuG418-derived CM affected Thy1.1 expression more when the feeders themselves were not present. This could be due to the feeders supporting themselves. Where feeders were present, those cells could uptake some of the soluble factors present in the CM, leaving a lower concentration of soluble factors for the OECs. This would mean that there were more soluble factors in the media when the feeders were not present. Fibroblasts participate in paracrine signalling [31,32], and therefore, it follows that if they are not present to receive the factors, the factors present can significantly assist other cells present.

p75NTR was present in all conditions at a similar level, with the exception of laminin with standard media, which was significantly lower at 4.0 ± 0.8 cells/mm^2. The presence of CM increased the expression of p75NTR on laminin (19.7 ± 7.6 cells/mm^2). An increase in p75NTR expression without the presence of the feeders would imply that there was some form of paracrine signalling occurring. Whether this was due to FGF2 expression or a combination of other factors would require a deeper examination of the components in the CM. Without the higher expression of Thy1.1, CM with laminin would be a promising condition to take forward. Thy1.1, historically an undesirable marker of fibroblast phenotype in the OEC field, negated the positive effect of the increase in putative OEC marker p75NTR as it implied a higher level of fibroblast impurities that would need to be removed from the culture.

The morphology of S100β-positive cells was analysed using an automated circularity macro in ImageJ (Figure 1N–Q). Numerical analysis revealed that there was no significant difference between the distributions for feeders versus feeders with CM, and both conditions had a left skew (skew = 0.4 and 0.5, respectively). The laminin condition had a skew value of 0.3 and a kurtosis value of −0.8. When CM was added to the laminin condition, the distribution was significantly different from the standard laminin condition (Kolmogorov–Smirnov, $p = 0.01$). The distribution with CM had a less flattened distribution (kurt = −0.3 compared to −0.8) and a more pronounced left shift (skew = 0.7 compared to 0.3). This was a stronger shift than the feeder conditions. This was significant as this was the condition with the most Thy1.1-positive cells. The most likely option was that spindle-shaped OECs were also expressing Thy1.1, as well as S100β. Co-labelling of these antibodies was not carried out in this study due to the lack of consistency experienced with the separate S100β antibody, which meant the Thy1.1 and S100β antibodies used were raised in the same species. The literature indicated that it is possible that OECs do co-express p75NTR and Thy1.1 [33–36] and Thy1.1 and GFAP [34], so the assumption that Thy1.1 is a contaminating cell marker may need to be re-examined. Due to the lack of putative OEC markers [37,38], these cells are poorly characterised, and this presents one of the major challenges of OEC research.

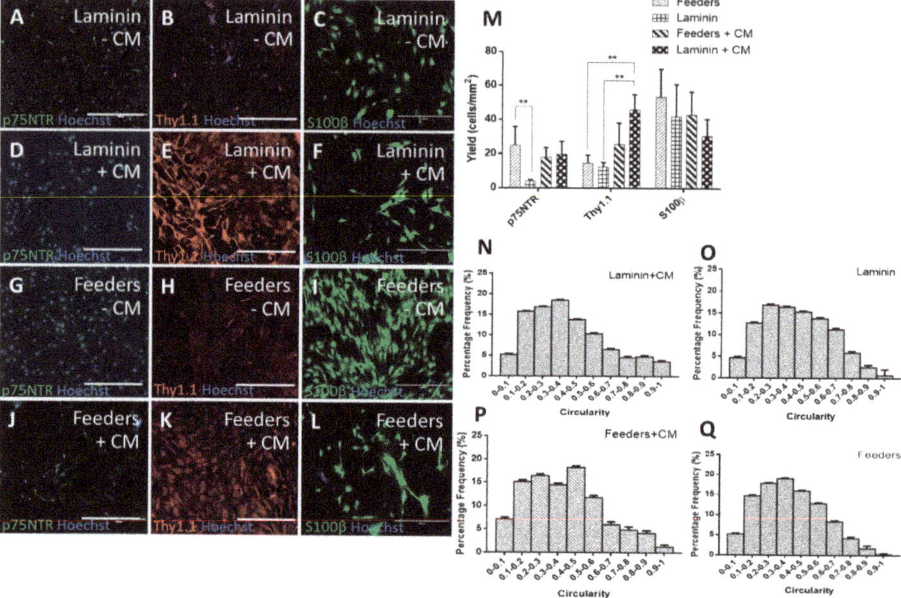

Figure 1. Fluorescent micrographs of primary rat olfactory mucosal cells (OMCs) cultured on laminin (**A–F**) and HuG418 feeders (**G–L**) in the presence (**D–F**, **J–L**) and absence (**A–C**, **G–I**) of HuG418 conditioned media and stained for olfactory ensheathing cell (OEC) biomarkers p75NTR and S100β and fibroblast marker Thy1.1. Positive cells were counted in ImageJ and calculated as the number of positive cells over the image area (**M**). OMCs cultured on laminin with standard media had the lowest yield for p75NTR, and the addition of conditioned media caused an upregulation of undesirable marker Thy1.1. Circularity was used as a measurement of morphology, and positive S100β cells were analysed for their circularity (**N–Q**). Cells on laminin with standard media and feeders with conditioned media gave more enlarged cell morphologies. The scale bars represent 400 µm. Data are the means ± SEM., n = 3. CM, conditioned media.

Although a differential adhesion step was used to remove rapidly adherent cells, leaving the slow attaching OECs in culture, the primary culture could still harbour contaminating cell types. This is demonstrated in Figure 2, where an entire well of OECs on feeders was imaged at 10× objective magnification (EVOS Life Technologies AMF4300) and the resulting images merged together using PanoramaPro2 in order to gain an understanding of the culture as a whole. It was observed that there were distinct areas of Thy1.1-positive colonies. When images were taken for counting, these images could yield data from 100% p75NTR-positive to 0% p75NTR-positive. This wide range of values resulted in a large standard deviation and therefore a large standard error of the mean across independent experiments (seen Figure 1M). These large errors were due to stochastic variability in the cultures due to the different cell subsets present, and these different cells could predominate in different regions of a single dish. From a cell manufacturing perspective, this created a significant challenge for creating a robust and well-characterised product.

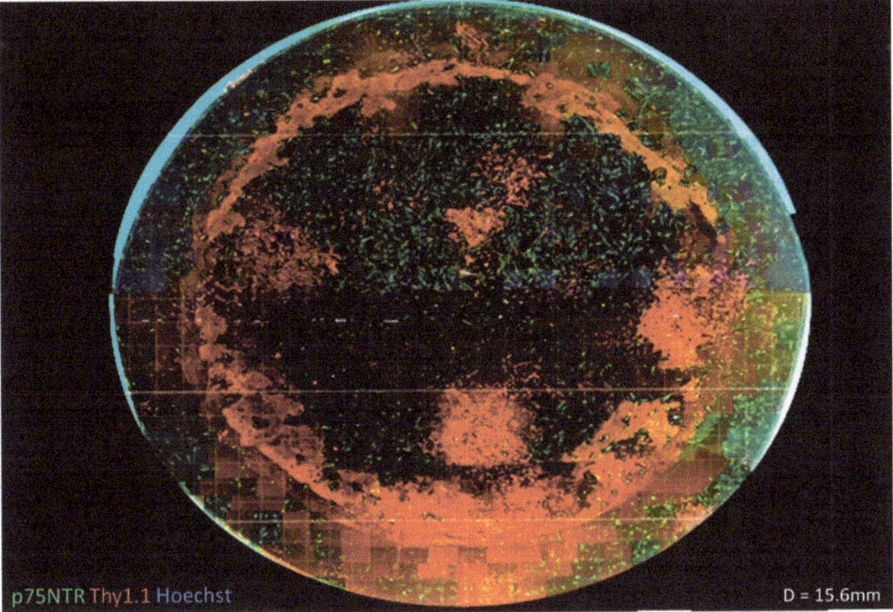

Figure 2. Assembled fluorescent micrograph of a cell population derived from rat olfactory mucosa cultured on a human feeder layer of olfactory fibroblasts. Images were taken at 100× total magnification in a 24 well plate (diameter = 15.6 mm) and stitched together using PanoramaPro2. It can be seen from this stitched image that there are definite populations present in the well plate. There are distinct areas of Thy1.1 (red) that do not appear to be dispersed with the areas of p75NTR (green). This variety in present populations explains the large error bars found in the cell counts. When multiple views were taken for counting, the purity could range from 0–100%, which relates to a large standard deviation.

After staining, there was also a red/yellow ring around the periphery of the plate that was not associated with any positive cells. On close inspection of the well plates used, it was found that the bottom of the wells was rarely smooth, and often, there were circular scratches, presumably as a result of the manufacturing process. It was postulated that the antibody interacted with this non-smooth surface or was trapped in ECM deposited as a consequence of cell responses to the surface scratches and gave the circular staining [39]. This pattern was observed in several images; however, it was not until the full well was imaged that it was noticed that it made a complete circle.

3.2. Comparison of Different Feeder Layers

One striking observation in the S100β (detecting OECs) and Fn (fibroblasts) staining condition comparing Ms3T3 and HuG418 feeders (Figure 3) was that the OECs did not appear to grow directly over the feeder layer. Instead, they grew close to the fibroblasts (Figure 3C,D). This placement made sense if the cells were gaining benefits from cell-to-cell contact. If paracrine factors were involved, the OECs would not necessarily border the fibroblasts so closely [40].

No differences in p75NTR expression were observed on the different feeder layers (Figure 3A,B,G) with yields of 15.9 ± 6.8 cells/mm^2 and 10.7 ± 5.3 cells/mm^2 on Ms3T3 and HuG418, respectively. This may indicate that the OECs were benefitting from general cell-to-cell contact, not specifically anatomically-matched HuG418 cells. However, as both lines were comprised of fibroblasts, there could also be a common paracrine factor involved. OECs have been well documented to grow poorly in isolation, and when a low yield is obtained, they do not proliferate [41–43]. Additionally, they have been observed to survive better in a culture with a mixed population (normally OECs and olfactory fibroblasts) [10,44–47]. This would also explain why the HuG418 CM did not have the anticipated beneficial impact on OECs' p75TNR expression as hoped as the cells themselves may be required. In the literature, it has been observed that there is a close relationship between OECs and olfactory fibroblasts. However, it does appear that this relationship is due to the physical contact as opposed to any paracrine factors present [48].

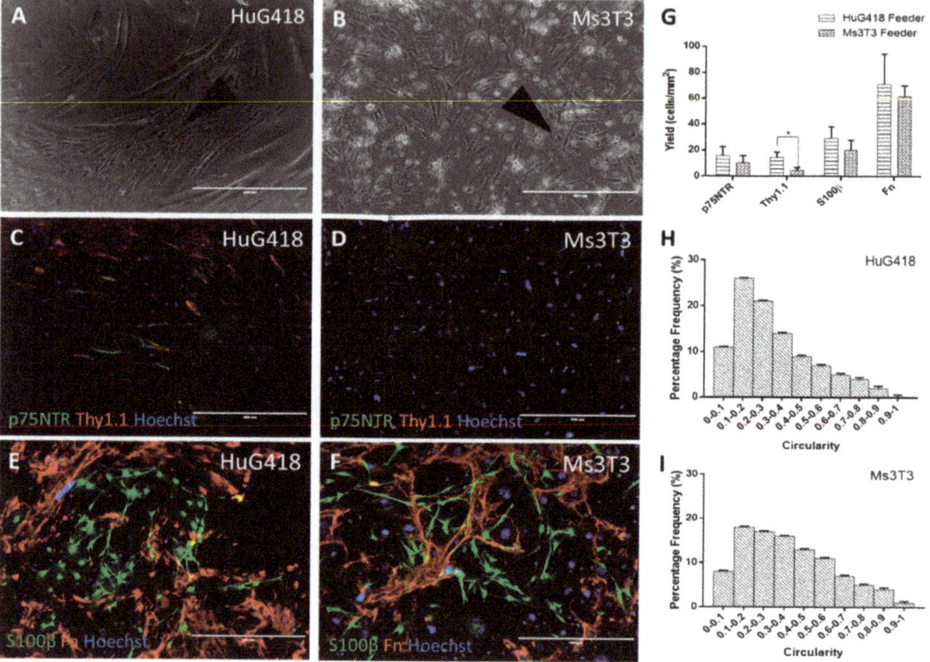

Figure 3. Cells were fixed after 14 days of culture and stained for OEC biomarkers p75NTR and S100β and fibroblast biomarkers Thy1.1 and Fn (**A–D**). Positive cells were quantified in ImageJ and graphed as the yield of positive cells in the image area (**E–G**). Circularity was used as a measurement of morphology, and positive S100β cells were analysed for their circularity (**H**,**I**). The scale bars represent 400 μm. Data are the means ± SEM, n = 3.

Thy1.1 expression was found to be significantly different between the two feeders (Kruskal–Wallis, $p = 0.01$). As observed in the ICC images and quantification, the presence of Ms3T3 downregulated the expression of Thy1.1 (yield of 14.6 ± 4.1 cells/mm^2 on HuG418 and 4.8 ± 2.1 cells/mm^2 on Ms3T3). This was important as it indicated that the Ms3T3 feeders were superior to the HuG418 when it came to supporting OEC growth and expression of p75NTR. OECs appeared to be supported by cell-to-cell contact and did not gain any benefit from paracrine factors. These results suggested that Ms3T3 feeders were superior to HuG418 cells as they resulted in fewer Thy1.1-positive cells whilst maintaining similar p75NTR expression levels.

These results, together with the results obtained with CM, indicated that OECs benefitted from cell-to-cell contact rather than paracrine signalling by HuG418 cells. Ms3T3 cells are commercially available as a GMP cell line (Sigma-Aldrich, UK). Therefore, their application as a feeder layer to enhance OEC phenotype during scalable manufacture is attractive to advance OECs towards the clinic. Potential safety concerns around the purity of the final clinical preparation could be addressed by purifying OECs with affinity-based removal of the mouse feeders. From this initial work with primary rat OECs, it was found that rat OECs have a more promising regenerative function when cultured with feeders. However, there was no difference between human olfactory mucosa-derived feeders and mouse 3T3 cells in their ability to support OEC marker expression.

3.3. Human OECs Behave Differently to Rat OECs in Culture

Studies in small animals are important to generate pre-clinical data and to gain an understanding of responses in physiologic systems before moving into humans. However, there are concerns that OECs in different animals behave differently [49–51], and animal cells and tissue do not accurately reflect the function or behaviour of human material. This is especially important when the length of the injury is considered. The gap to be bridged by the neurons in an SCI is much smaller in a rat compared to a human [50], and the ability of OECs to support neurons over this small distance may not be relevant on the human scale. Therefore, we generated conditionally immortalised cell lines from the human olfactory mucosa [26] and tested these under the same conditions in order to determine how translatable animal studies are.

ICC micrographs in Figure 4 revealed that Fn staining on laminin occurred in a completely different pattern from the other matrix conditions. In Ms3T3 and PLL conditions, the Fn deposition could be more easily traced back to the individual cells producing it. In contrast, on laminin, the staining pattern was more widely distributed. These distinct differences in the pattern were unexpected. Due to this staining pattern, cell counts and yield were not carried out for Fn due to the inaccuracy of determining the original cell responsible for the Fn. The reason for the different staining pattern on laminin is not known, and no references to similar occurrences could be found in the literature.

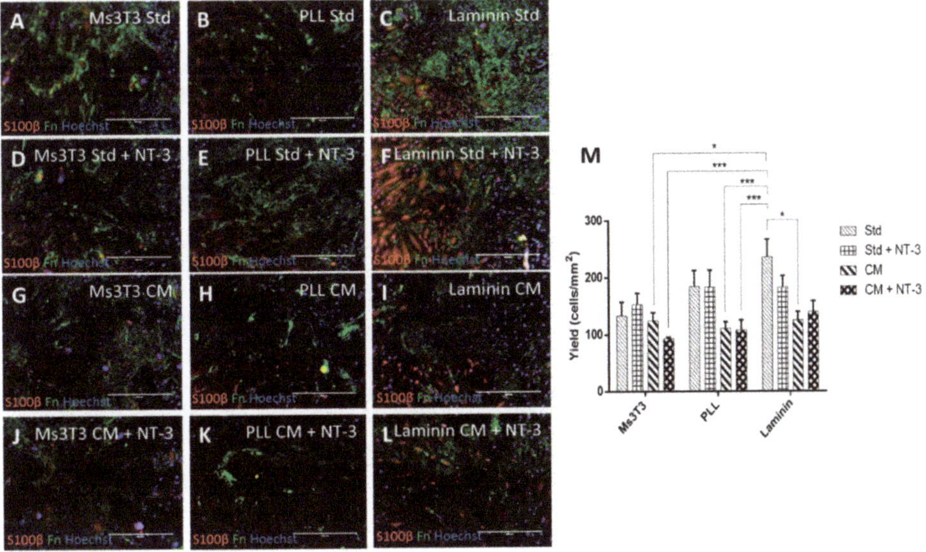

Figure 4. PA5 cells cultured on laminin or Ms3T3 feeders, in standard media or HuG418 conditioned media. Cells were cultured for five days and stained for S100β and Fn (**A–L**). S100β-positive cells were quantified in ImageJ and graphed as yield of positive cells in the image area (**M**). Conditioned media constantly gave lower yields of S100β, and laminin with standard media gave the highest yield. The scale bars represent 400μm. Data are the means ± SEM, n = 3.

S100β expression appeared to be more prominent in conditions with NT-3 compared with those conditions cultured in its absence and CM did not have any positive impact on S100β expression (Figure 4). From the three matrix conditions, it was found that the presence of Ms3T3 feeders was less beneficial for glial marker expression compared with PLL and laminin, which contrasted what was observed for rat OECs (yield of 133.7 ± 24.0 cells/mm^2 on Ms3T3 with standard media compared to 184.8 ± 27.7 cells/mm^2 and 235.9 ± 31.0 cells/mm^2 for PLL and laminin with standard media, respectively). This was not necessarily surprising, as it has been found that there are several differences between the rat and human olfactory system [52] including in vitro growth, spontaneous immortalisation, and morphology [53,54]. Additionally, in terms of implantation and isolation, the olfactory mucosa in rats is yellow, whereas in humans, it cannot be discerned from the respiratory tissue [55]. This leads to a higher level of non-OECs being present in the implant, which may or may not assist in regeneration. It does cause concern that any previous work that has been carried out in rat is not translatable to human scale up [24,49].

Another difference between the rat and human OECs is that the human OECs do not appear to benefit from the presence of NT-3. Although the study was not carried out as part of this work, previous studies have shown that NT-3 significantly increased the expression of glial markers in rat cells [27]. For our human cell line, the presence of NT-3 seemed to increase the level of expression slightly, but not significantly.

The differences that occurred between rat and human OECs in regard to neurotrophic factors and matrix preference showed that more in depth study and understanding of human OECs are necessary to be able to predict cell behaviour. It also indicated that care needs to be taken when translating results from rat to human OECs. Two human cell lines were investigated (data not shown from the second cell line), and they both followed the same pattern with laminin being the highest performing condition in terms of yield and significantly higher than most of the conditions with CM. Due to this, human OECs were carried forward into the future work to ensure the conditions investigated were relevant to the clinic.

3.4. Human OEC Co-culture with Neurons

Neurite outgrowth in cultured neurons is considered an indication of neuro-regenerative potential [56–60], and therefore, neurite outgrowth was quantified to compare the cultures. Neurite number and length were normalised to the number of neurites and neurons. Neurons were cultured in isolation to ensure any neurite extension we observed could be related to the presence of the OECs.

Cells co-cultured on laminin with standard media had a higher number of neurites with longer extensions (Figure 5). Cells co-cultured on laminin and feeders with CM did not perform as well as cells cultured with standard media. No difference was observed between the neuron-only conditions on the different matrices for all measurements. This indicated that the neurons were responding to the PA5s as opposed to the culture conditions (matrix and media).

The conditions cultured with OECs performed best in regard to average neurite length (Figure 5S, one-way ANOVA, Bonferroni post-hoc, $p < 0.001$). Although co-culture on laminin with standard media performed better than those co-cultured with CM (49.7 ± 3.7μm versus 40.4 ± 1.6 μm), this did not result in any significant difference.

When the neurite length per neuron was examined (Figure 5T), the co-culture on laminin with standard media showed significantly longer neurites per neuron (76.0 ± 11.2 μm) compared with every other condition (43.6 ± 5.5 μm on laminin with CM and 41.0 ± 10.6 μm on Ms3T3 with CM, one-way ANOVA, Bonferroni post-hoc, $p < 0.01$). All conditions without OECs performed similarly to each other (22.8 ± 2.4 μm, 17.8 ± 2.3 μm and 17.9 ± 1.1 μm on laminin with standard media, laminin with CM and Ms3T3 with CM, respectively).

The number of neurites extended by each neuron revealed a similar pattern (Figure 5U) where the co-culture on laminin with standard media produced more neurites per neuron (1.54 ± 0.20) than the majority of the other conditions (one-way ANOVA, Bonferroni post-hoc, $p < 0.001$) with the exception of co-culture on laminin with CM (1.08 ± 0.14). Collectively, these data showed that the matrix and media condition that gave the highest S100β expression in OECs resulted in the best neuronal growth support.

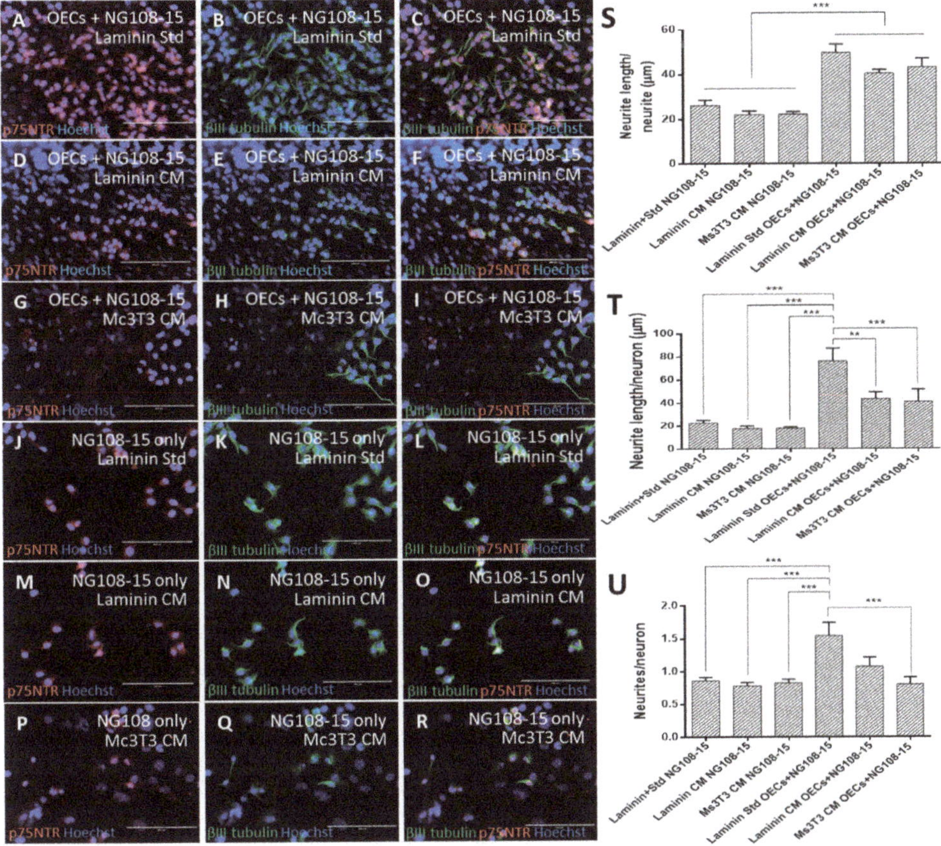

Figure 5. PA5 cells were co-cultured with NG108-15 neurons for five days under different matrix and media conditions. NG108-15 cells were also cultured in isolation to ensure that any improvement observed could be reliably attributed to the PA5 cells as opposed to the matrix and media conditions. After five days of NG108-15 culture, cells were fixed and stained for βIII-tubulin and p75NTR (**A–R**). The NeuronJ plugin in for Image J was used to quantify the neurite outgrowth, and measurements were made for neurite length/neurite (**S**), neurite length/neuron (**T**), and neurites/neurons (**U**). The scale bars represent 200 μm. Data are the means ± SEM, n = 3.

The formation of neurites from neurons is vital to the functionality and development of the nervous system [61]. From these results, it can be seen that co-culture on laminin with standard media was the condition that provided the most support to the development of neurons in regard to the longest average neurite. In the absence of co-culture, the development of neurons was not well supported, and as a result, shorter neurites were observed. Previously, it was found that laminin with standard media was a promising condition for OEC proliferation and S100β expression. This suggested that when OECs were expressing higher levels of S100β and that they were able to provide better support to neurons. When S100β expression was lower, the presence of the OECs was still able to benefit the neurons.

When Schwann cells support peripheral nerve repair, it has been observed that an upregulation of S100β leads to identification of so-called "reactive" Schwann cells, which are responsible for axonal sprouting [62]. If OECs followed a similar behavioural pattern, it could be that the conditions with higher S100β were more capable of allowing axonal sprouting and therefore resulted in increased neurite numbers.

It has been observed in several studies that laminin is an extracellular matrix that is able to stimulate rapid neurite growth and has been directly linked to neurite outgrowth in vitro [63–65]. Studies have also shown that enhanced neurite outgrowth and preferred attachment was observed when neurons from the CNS were plated on laminin [66], and it is thought that the presence of laminin starts to create a more permissible environment for axonal extension [67]. The hostile environment present after SCI is a key part of why nerve regeneration does not happen spontaneously [24,68,69]. There was no significant difference observed between neurons cultured in isolation on laminin and on Ms3T3 feeders. This suggests there is more to the interaction than the preference for laminin. The combined effect of the favourable matrix and support cells could explain why laminin with standard media was the highest performing condition.

CM collected from HuG418 had a lesser effect on the average neurite length per neuron compared to standard media. This may be initially related to the lower S100β expression in the OECs in this condition. It would indicate that the CM from HuG418 does not have any beneficial soluble paracrine factors for the neurons. This lack of factors and therefore interaction between these two populations is not necessarily unexpected as although studies have shown a benefit in transplantation with a mixed population of OECs and fibroblasts (Keyvan-Fouladi et al., 2003 [44], Ramón-Cueto et al., 2000 [10], Raisman and Li, 2007 [47], Teng et al., 2008 [46]), it has been under the understanding that the fibroblasts support the OECs, not the neurons. Fibroblasts have not been pursued as a cell therapy option for nerve regeneration, and they are not believed to have specific properties that enhance the function of neurons [70,71]. Therefore, standard media resulting in longer neurites than conditioned media can be seen as an expected output.

4. Conclusions

We established that the behaviour of rat and human OECs did not follow the same patterns, and therefore, caution needs to be engaged when using pre-clinical data that have been carried out using rat models. When focusing on human OECs, we found that out of the conditions investigated, laminin with NT-3 was the best condition for protein expression and neurite extension. It would be valuable to investigate the components of the fibroblast conditioned media to further understand what aspects resulted in Thy1 upregulation and how the population reacts when Thy1 positive cells are removed.

Author Contributions: Conceptualization, R.W. and I.W.; data curation, R.W. and P.D.; formal analysis, R.W. and P.D.; funding acquisition, I.W.; investigation, R.W. and P.D.; methodology, R.W. and P.D.; project administration, R.W. and I.W.; supervision, I.W.; visualization, R.W.; writing, original draft, R.W. and I.W.; writing, review and editing, R.W., P.D. and I.W. All authors have read and agreed to the published version of the manuscript.

Funding: This research was funded by the Biotechnology and Biological Sciences Research Council (BBSRC; Grant Number BB/K011154/1), the Bioprocessing Research Industry Club (BRIC), the UCL Overseas Research Scholarship, and the New Zealand Federation for Graduate Women (NZFGW).

Acknowledgments: We would like to thank the Biological Services Unit (BSU) at UCL for their help with culling the animals.

Conflicts of Interest: The authors declare no conflicting interest in this work.

References

1. World Health Organization. *International Perspectives on Spinal Cord Injury*; WHO: Geneva, Switzerland, 2013; p. 16.
2. National Spinal Cord Injury Center; National Spinal Cord Injury Statistical Center. *Fact Sheet*; National Spinal Cord Injury Center; National Spinal Cord Injury Statistical Center: Glasgow, UK, 2006.
3. Willerth, S.M.; Sakiyama-Elbert, S.E. Cell therapy for spinal cord regeneration. *Adv. Drug Deliv.Rev.* **2008**, *60*, 263–276. [CrossRef]

4. Gwak, Y.S.; Kang, J.; Unabia, G.C.; Hulsebosch, C.E. Spatial and temporal activation of spinal glial cells: Role of gliopathy in central neuropathic pain following spinal cord injury in rats. *Exp. Neurol.* **2012**, *234*, 362–372. [CrossRef] [PubMed]
5. Fawcett, J.W.; Asher, R.A. The glial scar and central nervous system repair. *Brain Res. Bull.* **1999**, *49*, 377–391. [CrossRef]
6. Ronaghi, M.; Erceg, S.; Moreno-Manzano, V.; Stojkovic, M. Challenges of stem cell therapy for spinal cord injury: Human embryonic stem cells, endogenous neural stem cells, or induced pluripotent stem cells? *Stem Cells* **2010**, *28*, 93–99. [CrossRef] [PubMed]
7. Kigerl, K.A.; de Rivero Vaccari, J.P.; Dietrich, W.D.; Popovich, P.G.; Keane, R.W. Pattern recognition receptors and central nervous system repair. *Exp. Neurol.* **2014**, *258*, 5–16. [CrossRef] [PubMed]
8. Yang, H.; He, B.R.; Hao, D.J. Biological Roles of Olfactory Ensheathing Cells in Facilitating Neural Regeneration: A Systematic Review. *Mol. Neurobiol.* **2015**, *51*, 168–179. [CrossRef]
9. Fraher, J.P. The transitional zone and CNS regeneration. *J. Anat.* **2000**, *196*, 137–158. [CrossRef]
10. Ramón-Cueto, A.; Cordero, M.I.; Santos-Benito, F.F.; Avila, J. Functional Recovery of Paraplegic Rats and Motor Axon Regeneration in Their Spinal Cords by Olfactory Ensheathing Glia. *Neuron* **2000**, *25*, 425–435. [CrossRef]
11. Kawaja, M.D.; Boyd, J.G.; Smithson, L.J.; Jahed, A.; Doucette, R. Technical strategies to isolate olfactory ensheathing cells for intraspinal implantation. *J. Neurotrauma* **2009**, *26*, 155–177. [CrossRef]
12. Thuret, S.; Moon, L.D.F.; Gage, F.H. Therapeutic interventions after spinal cord injury. *Nat. Rev. Neurosci.* **2006**, *7*, 628–643. [CrossRef]
13. Orbay, H.; Little, C.J.; Lankford, L.; Olson, C.A.; Sahar, D.E. The Key Components of Schwann Cell-like Differentiation Medium and their Effects on Gene Expression Pattern of Adipose-Derived Stem Cells. *Ann. Plast. Surg.* **2015**, *74*, 584–588. [CrossRef]
14. Tong, L.; Ji, L.; Wang, Z.; Tong, X.; Zhang, L.; Sun, X. Differentiation of neural stem cells into Schwann-like cells in vitro. *Biochem. Biophys. Res. Commun.* **2010**, *401*, 592–597. [CrossRef] [PubMed]
15. Niapour, A.; Karamali, F.; Karbalaie, K.; Kiani, A.; Mardani, M.; Nasr-Esfahani, M.H.; Baharvand, H. Novel method to obtain highly enriched cultures of adult rat Schwann cells. *Biotechnol. Lett.* **2010**, *32*, 781–786. [CrossRef] [PubMed]
16. Chen, Y.; Zeng, J.; Chen, Y.; Wang, X.; Yao, G.; Wang, W.; Qi, W.; Kong, K. Multiple roles of the p75 neurotrophin receptor in the nervous system. *J. Int. Med. Res.* **2009**, *37*, 281–288. [CrossRef] [PubMed]
17. Chen, C.R.; Kachramanoglou, C.; Li, D.Q.; Andrews, P.; Choi, D. Anatomy and Cellular Constituents of the Human Olfactory Mucosa: A Review. *J. Neurol. Surg. Part B* **2014**, *75*, 293–300. [CrossRef]
18. Chong, Z.Z. S100B raises the alert in subarachnoid hemorrhage. *Rev. Neurosci.* **2016**, *27*, 745–759. [CrossRef]
19. Wu, X.; Bolger, W.E.; Anders, J.J. Fibroblasts isolated from human middle turbinate mucosa cause neural progenitor cells to differentiate into glial lineage cells. *PLoS ONE* **2013**, *8*, e76926. [CrossRef]
20. Ebel, C.; Brandes, G.; Radtke, C.; Rohn, K.; Wewetzer, K. Clonal In Vitro Analysis of Neurotrophin Receptor p75-Immunofluorescent Cells Reveals Phenotypic Plasticity of Primary Rat Olfactory Ensheathing Cells. *Neurochem. Res.* **2013**, *38*, 1078–1087. [CrossRef]
21. Ramon-Cueto, A.; Perez, J.; Nieto-Sampedro, M. In vitro enfolding of olfactory neurites by p75 NGF receptor positive ensheathing cells from adult rat olfactory bulb. *Eur. J. Neurosci.* **1993**, *5*, 1172–1180. [CrossRef]
22. Rath, N.; Balain, B. Spinal cord injury-The role of surgical treatment for neurological improvement. *J. Clin. Orthop. Trauma* **2017**, *8*, 99–102. [CrossRef]
23. Raisman, G.; Carlstedt, T.; Choi, D.; Li, Y. Clinical prospects for transplantation of OECs in the repair of brachial and lumbosacral plexus injuries: Opening a door. *Exp. Neurol.* **2011**, *229*, 168–173. [CrossRef] [PubMed]
24. Ahuja, C.S.; Wilson, J.R.; Nori, S.; Kotter, M.R.N.; Druschel, C.; Curt, A.; Fehlings, M.G. Traumatic spinal cord injury. *Nat. Rev. Dis. Primers* **2017**, *3*, 17018. [CrossRef] [PubMed]
25. Bramlett, H.M.; Dietrich, W.D. Progressive damage after brain and spinal cord injury: Pathomechanisms and treatment strategies. In *Progress in Brain Research*; John, T.W., Andrew, I.R.M., Eds.; Elsevier: Amsterdam, The Netherlands, 2007; Volume 161, pp. 125–141.
26. Santiago-Toledo, G.; Georgiou, M.; dos Reis, J.; Roberton, V.H.; Valinhas, A.; Wood, R.C.; Phillips, J.B.; Mason, C.; Li, D.; Li, Y.; et al. Generation of c-MycERTAM-transduced human late-adherent olfactory mucosa cells for potential regenerative applications. *Sci. Rep.* **2019**, *9*, 13190. [CrossRef] [PubMed]

27. Georgiou, M.; Reis, J.N.D.; Wood, R.; Esteban, P.P.; Roberton, V.; Mason, C.; Li, D.; Li, Y.; Choi, D.; Wall, I. Bioprocessing strategies to enhance the challenging isolation of neuro-regenerative cells from olfactory mucosa. *Sci. Rep.* **2018**, *8*, 14440. [CrossRef]
28. Stationery Office. *Animals (Scientific Procedures) Act 1986: Guidance on the Operation of the Animals (Scientific Procedures) Act 1986*; Stationery Office: London, UK, 2000.
29. Pollock, K.; Stroemer, P.; Patel, S.; Stevanato, L.; Hope, A.; Miljan, E.; Dong, Z.; Hodges, H.; Price, J.; Sinden, J.D. A conditionally immortal clonal stem cell line from human cortical neuroepithelium for the treatment of ischemic stroke. *Exp. Neurol.* **2006**, *199*, 143–155. [CrossRef]
30. Meijering, E.; Jacob, M.; Sarria, J.C.; Steiner, P.; Hirling, H.; Unser, M. Design and validation of a tool for neurite tracing and analysis in fluorescence microscopy images. *Cytom. Part A* **2004**, *58*, 167–176. [CrossRef]
31. Alberts, B.; Johnson, A.; Lewis, J.; Raff, M.; Roberts, K.; Walter, P. *Molecular Biology of the Cell*, 4th ed.; Garland Science: New York, NY, USA, 2002.
32. Wojtowicz, A.M.; Oliveira, S.; Carlson, M.W.; Zawadzka, A.; Rousseau, C.F.; Baksh, D. The importance of both fibroblasts and keratinocytes in a bilayered living cellular construct used in wound healing. *Wound Rep. Regen.* **2014**, *22*, 246–255. [CrossRef]
33. Hayat, S.; Thomas, A.; Afshar, F.; Sonigra, R.; Wigley, C.B. Manipulation of olfactory ensheathing cell signaling mechanisms: Effects on their support for neurite regrowth from adult CNS neurons in coculture. *Glia* **2003**, *44*, 232–241. [CrossRef]
34. Nash, H.H.; Borke, R.C.; Anders, J.J. New method of purification for establishing primary cultures of ensheathing cells from the adult olfactory bulb. *Glia* **2001**, *34*, 81–87. [CrossRef]
35. Nash, H.H.; Borke, R.C.; Anders, J.J. Ensheathing cells and methylprednisolone promote axonal regeneration and functional recovery in the lesioned adult rat spinal cord. *J. Neurosci.* **2002**, *22*, 7111–7120. [CrossRef]
36. Sonigra, R.J.; Brighton, P.C.; Jacoby, J.; Hall, S.; Wigley, C.B. Adult rat olfactory nerve ensheathing cells are effective promoters of adult central nervous system neurite outgrowth in coculture. *Glia* **1999**, *25*, 256–269. [CrossRef]
37. Ulrich, R.; Imbschweiler, I.; Kalkuhl, A.; Lehmbecker, A.; Ziege, S.; Kegler, K.; Becker, K.; Deschl, U.; Wewetzer, K.; Baumgartner, W. Transcriptional profiling predicts overwhelming homology of Schwann cells, olfactory ensheathing cells, and Schwann cell-like glia. *Glia* **2014**, *62*, 1559–1581. [CrossRef] [PubMed]
38. Barbacid, M. The Trk family of neurotrophin receptors. *J. Neurobiol.* **1994**, *25*, 1386–1403. [CrossRef] [PubMed]
39. Stephens, P.; Genever, P.G.; Wood, E.J.; Raxworthy, M.J. Integrin receptor involvement in actin cable formation in an in vitro model of events associated with wound contraction. *Int. J. Biochem. Cell Biol.* **1997**, *29*, 121–128. [CrossRef]
40. Li, Y.; Sauve, Y.; Li, D.; Lund, R.D.; Raisman, G. Transplanted olfactory ensheathing cells promote regeneration of cut adult rat optic nerve axons. *J. Neurosci.* **2003**, *23*, 7783–7788. [CrossRef] [PubMed]
41. Lakatos, A.; Smith, P.M.; Barnett, S.C.; Franklin, R.J.M. Meningeal cells enhance limited CNS remyelination by transplanted olfactory ensheathing cells. *Brain* **2003**, *126*, 598–609. [CrossRef] [PubMed]
42. Blumenthal, J.; Cohen-Matsliah, S.I.; Levenberg, S. Olfactory Bulb-Derived Cells Seeded on 3D Scaffolds Exhibit Neurotrophic Factor Expression and Pro-Angiogenic Properties. *Tissue Eng. Part A* **2013**, *19*, 2284–2291. [CrossRef]
43. Kachramanoglou, C.; Law, S.; Andrews, P.; Li, D.; Choi, D. Culture of Olfactory Ensheathing Cells for Central Nerve Repair: The Limitations and Potential of Endoscopic Olfactory Mucosal Biopsy. *Neurosurgery* **2013**, *72*, 170–178. [CrossRef]
44. Keyvan-Fouladi, N.; Raisman, G.; Li, Y. Functional repair of the corticospinal tract by delayed transplantation of olfactory ensheathing cells in adult rats. *J. Neurosci.* **2003**, *23*, 9428–9434. [CrossRef]
45. Tabakow, P.; Raisman, G.; Fortuna, W.; Czyz, M.; Huber, J.; Li, D.; Szewczyk, P.; Okurowski, S.; Miedzybrodzki, R.; Czapiga, B.; et al. Functional Regeneration of Supraspinal Connections in a Patient with Transected Spinal Cord Following Transplantation of Bulbar Olfactory Ensheathing Cells with Peripheral Nerve Bridging. *Cell Trans.* **2014**, *23*, 1631–1655. [CrossRef]
46. Teng, X.; Nagata, I.; Li, H.-P.; Kimura-Kuroda, J.; Sango, K.; Kawamura, K.; Raisman, G.; Kawano, H. Regeneration of nigrostriatal dopaminergic axons after transplantation of olfactory ensheathing cells and fibroblasts prevents fibrotic scar formation at the lesion site. *J. Neurosci. Res.* **2008**, *86*, 3140–3150. [CrossRef] [PubMed]

47. Raisman, G.; Li, Y. Repair of neural pathways by olfactory ensheathing cells. *Nat. Rev. Neurosci.* **2007**, *8*, 312–319. [CrossRef] [PubMed]
48. Jani, H.R.; Raisman, G. Ensheathing cell cultures from the olfactory bulb and mucosa. *Glia* **2004**, *47*, 130–137. [CrossRef]
49. Wewetzer, K.; Radtke, C.; Kocsis, J.; Baumgartner, W. Species-specific control of cellular proliferation and the impact of large animal models for the use of olfactory ensheathing cells and Schwann cells in spinal cord repair. *Exp. Neurol.* **2011**, *229*, 80–87. [CrossRef]
50. Li, Y.; Decherchi, P.; Raisman, G. Transplantation of olfactory ensheathing cells into spinal cord lesions restores breathing and climbing. *J. Neurosci.* **2003**, *23*, 727–731. [CrossRef]
51. Hahn, C.-G.; Han, L.-Y.; Rawson, N.E.; Mirza, N.; Borgmann-Winter, K.; Lenox, R.H.; Arnold, S.E. In vivo and in vitro neurogenesis in human olfactory epithelium. *J. Comp. Neurol.* **2005**, *483*, 154–163. [CrossRef]
52. Krudewig, C.; Deschl, U.; Wewetzer, K. Purification and in vitro characterization of adult canine olfactory ensheathing cells. *Cell Tissue Res.* **2006**, *326*, 687–696. [CrossRef]
53. Omar, M.; Hansmann, F.; Kreutzer, R.; Kreutzer, M.; Brandes, G.; Wewetzer, K. Cell Type- and Isotype-Specific Expression and Regulation of β-Tubulins in Primary Olfactory Ensheathing Cells and Schwann Cells In Vitro. *Neurochem. Res.* **2013**, *38*, 981–988. [CrossRef]
54. Rubio, M.-P.; Munoz-Quiles, C.; Ramon-Cueto, A. Adult olfactory bulbs from primates provide reliable ensheathing glia for cell therapy. *Glia* **2008**, *56*, 539–551. [CrossRef]
55. Bianco, J.I.; Perry, C.; Harkin, D.G.; Mackay-Sim, A.; Feron, F. Neurotrophin 3 promotes purification and proliferation of olfactory ensheathing cells from human nose. *Glia* **2004**, *45*, 111–123. [CrossRef]
56. Lozano, A.M.; Schmidt, M.; Roach, A. A convenient in vitro assay for the inhibition of neurite outgrowth by adult mammalian CNS myelin using immortalized neuronal cells. *J. Neurosci. Methods* **1995**, *63*, 23–28. [CrossRef]
57. Encinas, M.; Iglesias, M.; Liu, Y.; Wang, H.; Muhaisen, A.; Cena, V.; Gallego, C.; Comella, J.X. Sequential treatment of SH-SY5Y cells with retinoic acid and brain-derived neurotrophic factor gives rise to fully differentiated, neurotrophic factor-dependent, human neuron-like cells. *J. Neurochem.* **2000**, *75*, 991–1003. [CrossRef]
58. Lehmann, M.; Fournier, A.; Selles-Navarro, I.; Dergham, P.; Sebok, A.; Leclerc, N.; Tigyi, G.; McKerracher, L. Inactivation of Rho signaling pathway promotes CNS axon regeneration. *J. Neurosci.* **1999**, *19*, 7537–7547. [CrossRef]
59. Fournier, A.E.; Gould, G.C.; Liu, B.P.; Strittmatter, S.M. Truncated soluble Nogo receptor binds Nogo-66 and blocks inhibition of axon growth by myelin. *J. Neurosci.* **2002**, *22*, 8876–8883. [CrossRef]
60. Simpson, P.B.; Bacha, J.I.; Palfreyman, E.L.; Woollacott, A.J.; McKernan, R.M.; Kerby, J. Retinoic acid-evoked differentiation of neuroblastoma cells predominates over growth factor stimulation: An automated image capture and quantitation approach to neuritogenesis. *Anal. Biochem.* **2001**, *298*, 163–169. [CrossRef]
61. Lai, P.-L.; Naidu, M.; Sabaratnam, V.; Kah Hui, W.; David, P.; Kuppusamy, U.R.; Abdullah, N.; Abd Malek, S.N. Neurotrophic Properties of the Lion's Mane Medicinal Mushroom, Hericium erinaceus (Higher Basidiomycetes) from Malaysia. *Int. J. Med. Mushrooms* **2013**, *15*, 539–554. [CrossRef] [PubMed]
62. Son, Y.-J.; Trachtenberg, J.T.; Thompson, W.J. Schwann cells induce and guide sprouting and reinnervation of neuromuscular junctions. *Trends Neurosci.* **1996**, *19*, 280–285. [CrossRef]
63. Menager, C.; Arimura, N.; Fukata, Y.; Kaibuchi, K. PIP3 is involved in neuronal polarization and axon formation. *J. Neurochem.* **2004**, *89*, 109–118. [CrossRef] [PubMed]
64. Chen, Z.-L.; Yu, W.-M.; Strickland, S. Peripheral regeneration. *Annu. Rev. Neurosci.* **2007**, *30*, 209–233. [CrossRef]
65. Kuhn, T.B.; Schmidt, M.F.; Kater, S.B. Laminin and Fibronectin Guideposts signal sustained but opposite effects to passing growth cones. *Neuron* **1995**, *14*, 275–285. [CrossRef]
66. Liesi, P.; Dahl, D.; Vaheri, A. Neurons cultured from developing rat brain attach and spread preferentially to laminin. *J. Neurosci. Res.* **1984**, *11*, 241–251. [CrossRef] [PubMed]
67. di Summa, P.G.; Kalbermatten, D.F.; Raffoul, W.; Terenghi, G.; Kingham, P.J. Extracellular Matrix Molecules Enhance the Neurotrophic Effect of Schwann Cell-Like Differentiated Adipose-Derived Stem Cells and Increase Cell Survival Under Stress Conditions. *Tissue Eng. Part A* **2013**, *19*, 368–379. [CrossRef] [PubMed]
68. Hagg, T.; Oudega, M. Degenerative and spontaneous regenerative processes after spinal cord injury. *J. Neurotrauma* **2006**, *23*, 264–280. [CrossRef] [PubMed]

69. Franssen, E.H.P.; de Bree, F.M.; Verhaagen, J. Olfactory ensheathing glia: Their contribution to primary olfactory nervous system regeneration and their regenerative potential following transplantation into the injured spinal cord. *Brain Res. Rev.* **2007**, *56*, 236–258. [CrossRef]
70. Pizzi, M.A.; Crowe, M.J. Transplantation of fibroblasts that overexpress matrix metalloproteinase-3 into the site of spinal cord injury in rats. *J. Neurotrauma* **2006**, *23*, 1750–1765. [CrossRef]
71. Kisselbach, L.; Merges, M.; Bossie, A.; Boyd, A. CD90 Expression on human primary cells and elimination of contaminating fibroblasts from cell cultures. *Cytotechnology* **2009**, *59*, 31–44. [CrossRef]

 *bioengineering*

Communication

Challenges and Solutions for Commercial Scale Manufacturing of Allogeneic Pluripotent Stem Cell Products

Brian Lee [1,*], Breanna S. Borys [2], Michael S. Kallos [2], Carlos A. V. Rodrigues [3], Teresa P. Silva [3] and Joaquim M. S. Cabral [3]

1 PBS Biotech, Inc., Camarillo, CA 93012, USA
2 Department of Chemical and Petroleum Engineering, Schulich School of Engineering, University of Calgary, 2500 University Dr. NW, Calgary, AB T2N 1N4, Canada; bsborys@ucalgary.ca (B.S.B.); mskallos@ucalgary.ca (M.S.K.)
3 Ibb—Institute for Bioengineering and Biosciences and Department of Bioengineering, Instituto Superior Técnico, Universidade de Lisboa, Lisboa 1049-001, Portugal; carlos.rodrigues@tecnico.ulisboa.pt (C.A.V.R.); teresasilva@tecnico.ulisboa.pt (T.P.S.); joaquim.cabral@tecnico.ulisboa.pt (J.M.S.C.)
* Correspondence: blee@pbsbiotech.com

Received: 29 February 2020; Accepted: 26 March 2020; Published: 28 March 2020

Abstract: Allogeneic cell therapy products, such as therapeutic cells derived from pluripotent stem cells (PSCs), have amazing potential to treat a wide variety of diseases and vast numbers of patients globally. However, there are various challenges related to the manufacturing of PSCs in large enough quantities to meet commercial needs. This manuscript addresses the challenges for the process development of PSCs production in a bioreactor, and also presents a scalable bioreactor technology that can be a possible solution to remove the bottleneck for the large-scale manufacturing of high-quality therapeutic cells derived from PSCs.

Keywords: allogeneic cell therapy; induced pluripotent stem cell; human embryonic stem cell; cell aggregate; expansion; differentiation; scalable manufacturing; scale up; single-use bioreactor; Vertical-Wheel; U-shaped vessel; computational fluid dynamics; shear stress; turbulent energy dissipation rates; homogeneous hydrodynamic environment

---

1. Introduction

With their potential to cure a wide variety of disease indications and address vast patient populations, allogeneic cell therapies derived from pluripotent stem cells (PSCs) are poised to revolutionize therapeutic medicines [1,2]. However, 2D planar manufacturing technologies that have been commonly used for small scale R&D and early stage clinical trials are inadequate and cost-prohibitive for production at the larger scales required for late-stage clinical trials and commercial manufacturing [3]. Single-use bioreactors are widely recognized as a feasible manufacturing solution but need to be optimized in order to meet the unique process requirements of PSCs [4]. One of the critical challenges for future success in commercializing allogeneic cell therapy products is establishing a scalable manufacturing technology that can reliably reproduce the yield and quality of PSC-derived products generated from small-scale R&D at larger scales sufficient for commercial manufacturing [5].

PSCs are mortal human cells that include specific cell types such as human embryonic stem cells (hESCs) and induced pluripotent stem cells (iPSCs). When cultivated in a 2D planar vessel, PSCs attach to a surface substrate and grow as a monolayer. In contrast, starting from single cells or small clumps in suspension bioreactors, PSCs that come into contact will naturally clump together to form spherical cell aggregates [6]. The formation of cell aggregates is required not only for the cell expansion phase but also for subsequent differentiation, which can be a multi-step process that directs the pluripotent cells to turn into a final target cell type for treating a particular disease.

Depending on a particular cell expansion and differentiation process, there is an optimal range of spherical cell aggregate sizes that can maximize the efficiency and production yield of expansion and differentiation processes in a bioreactor [7–9]. If a PSC aggregate becomes too large, nutrients and the growth or differentiation factors may be unable to evenly diffuse from the aggregate surface into its center, leading to unwanted cell death or heterogeneous cell populations during expansion or differentiation [10,11]. PSC aggregates that are too small may result in less efficient cell expansion and differentiation, thus lowering the yields of final target cells [12].

There are varying published examples of optimal cell aggregate sizes for different cell types and cell culture process steps. One experiment using hESCs showed that maximum viability and minimal cell apoptosis during expansion was achieved when average aggregate diameter was 300 μm after 7 days of expansion, while apoptosis (with the majority of necrotic cells in the centers of aggregates) peaked by day 14 if the average diameter reached 500 μm [13]. In another experiment, hESC aggregates with diameters of 400 μm detrimentally had half the concentration of oxygen in their centers compared to aggregates of 200 μm [10]. In a different experiment focusing on hESC differentiation, an average diameter of 450 μm was ideal for cardiogenesis, whereas diameters between 150–300 μm were ideal for endothelial cell differentiation [14]. For two different experiments involving hPSCs, an average aggregate diameter of 139 ± 26 μm was optimal for initiating differentiation into neural cells in one experiment [15], while an average diameter of 130 ± 40 μm was optimal for endoderm induction in the other experiment [16]. As these various experiments indicate, optimal cell aggregate size is highly dependent on factors such as cell type and expansion or differentiation process steps.

While PSC aggregate formation can be influenced by variables such as cell proliferation rate, cell–cell adhesion strength, and cell packing density, the hydrodynamic environment inside a bioreactor, which is created by the impeller used to continually mix the liquid media, also has a significant impact on determining aggregate size and, ultimately, cell viability [17]. The hydrodynamic environment is characterized by the two parameters of fluid shear stress and turbulent energy dissipation rate. Both of these parameters are inversely correlated with the average size of PSC aggregates: higher levels of shear stress and turbulent energy dissipation rate result in smaller sized aggregates, and vice versa. After the initial seeding of single cells or small preformed aggregates into a bioreactor, collisions due to the hydrodynamic environment will facilitate aggregate growth, either through the addition of single cells onto existing aggregates or the fusion of smaller aggregates into larger ones. At sufficiently higher agitation rates, the increased levels of shear stress will promote the breakage of loosely attached or temporarily agglomerated larger aggregates and thus limit their maximum possible size [18].

Fluid mixing in a bioreactor using a traditional horizontal-blade impeller creates a significantly uneven hydrodynamic environment. The highest levels of shear stress and turbulent energy dissipation rates will be near the tips of the rapidly spinning impeller, with decreasing gradients of these hydrodynamic parameters as the distance from the impeller increases [19,20]. Such a wide range of shear stress levels and energy dissipation rates in a horizontal-blade impeller bioreactor results in a non-uniform hydrodynamic environment, which in turn results in a wide variation in PSC aggregate sizes. In this scenario the size of cell aggregates will vary and the range of aggregate diameters in suspension becomes broad, which ultimately results in an inconsistent yield and quality of PSCs during expansion and differentiation steps. This variation in PSC aggregate size becomes more pronounced at larger sizes of horizontal-blade impeller bioreactors.

2. Scalable Bioreactor Technology as Manufacturing Solution

In contrast, the innovative Vertical-Wheel™ impeller, in conjunction with a distinct U-shaped vessel, provides a significantly more homogeneous hydrodynamic environment in Vertical-Wheel bioreactors. Computational fluid dynamics (CFD) analysis shows consistently low shear stress levels on all surfaces of the Vertical-Wheel impeller, as well as a narrow range distribution of turbulent energy dissipation rates throughout the U-shaped vessel (Figure 1) [21].

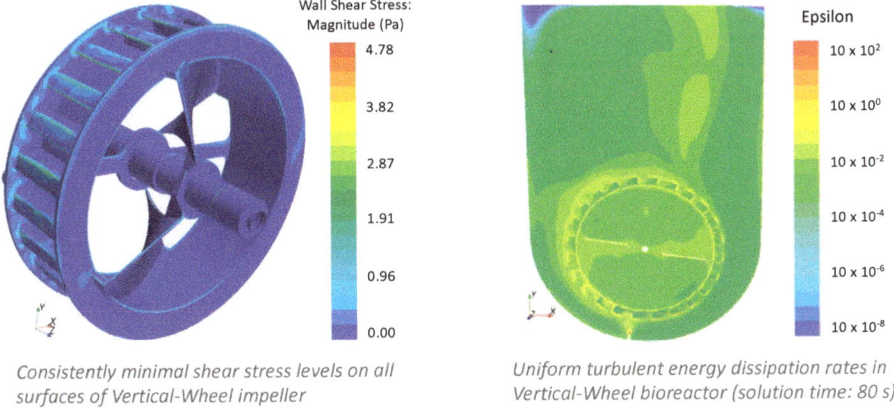

Figure 1. Computational fluid dynamics (CFD) analyses of shear stress on the surface of vertical-wheel impeller and range of turbulent energy dissipation rates in U-shaped vessel.

Fluid mixing using the Vertical-Wheel results in a very low variation in these two hydrodynamic conditions while still achieving the complete and continual suspension of PSC aggregates. The homogeneous hydrodynamic environment will result in a much tighter distribution of PSC aggregate size (Figure 2A). As a result, PSC aggregates in a Vertical-Wheel bioreactor will consistently have spherical shapes with similar diameters.

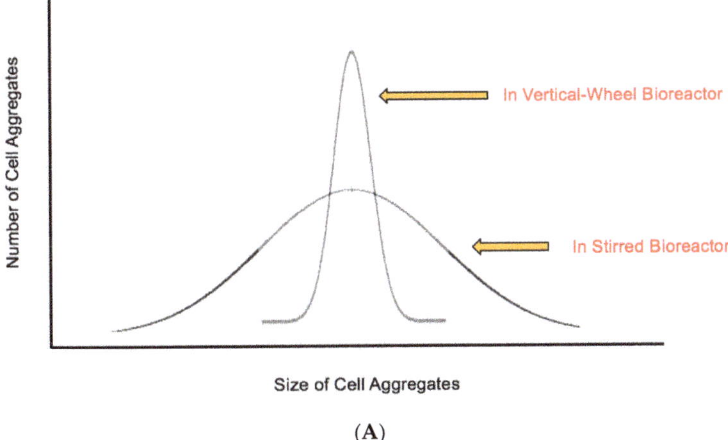

(A)

Figure 2. *Cont.*

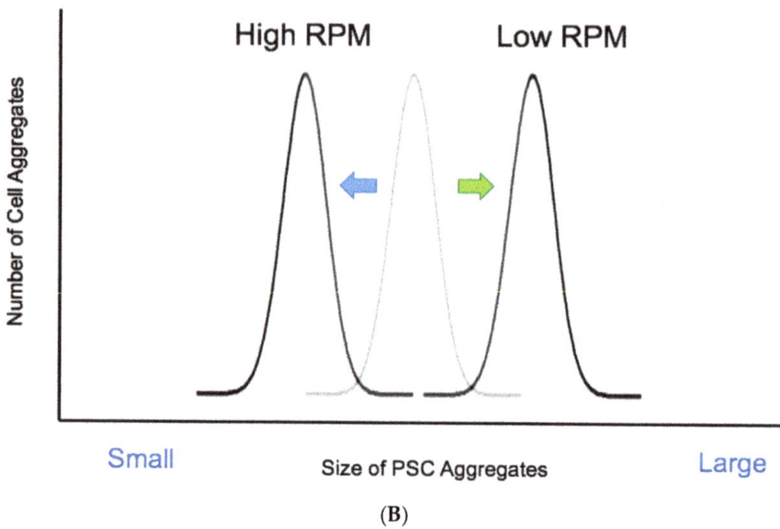

(B)

Figure 2. (**A**) Projected distribution of cell aggregate sizes in Vertical-Wheel bioreactor (homogeneous) vs. horizontally stirred bioreactor (non-homogeneous) hydrodynamic environments. (**B**) inverse correlation between average pluripotent stem cells (PSC) aggregate diameter and agitation rate in Vertical-Wheel bioreactors.

In addition, the average diameter of PSC aggregates can be controlled by simply adjusting the agitation rate of the Vertical-Wheel impeller (Figure 2B). As agitation rate increases, turbulent energy dissipation rates and shear stress levels also increase while maintaining a homogeneous hydrodynamic environment, resulting in smaller average aggregate diameters and still narrow size distributions. The inverse is also true, with average aggregate sizes becoming larger as agitation rate is lowered. Starting with a relatively low agitation rate after seeding can promote initial aggregate spheroid formation. The agitation rate could later be increased to a speed that completely suspends larger particles and forms optimally sized cell aggregates, while also preventing unwanted fusion or agglomeration. Furthermore, the homogeneous hydrodynamic environment is maintained as bioreactor volume increases. Being able to predict the mixing properties and hydrodynamic environment at larger scales based on process development done at a small scale will be enormously beneficial for establishing a scalable manufacturing process for PSCs.

3. Independent Biological Performance Data from Collaborators

Biological studies in Vertical-Wheel bioreactors have confirmed the inverse correlation between agitation rate and average size for iPSC aggregates (Figure 3A). The uniformity of cell aggregate sizes at different agitation rates in small-scale Vertical-Wheel bioreactors, compared to those produced in horizontal spinner flasks, was also confirmed (Figure 3B) [22]. The observed narrow size distributions and uniform morphology also indicate that unwanted agglomeration into overly large or irregular spheroids is minimized across various agitation rates.

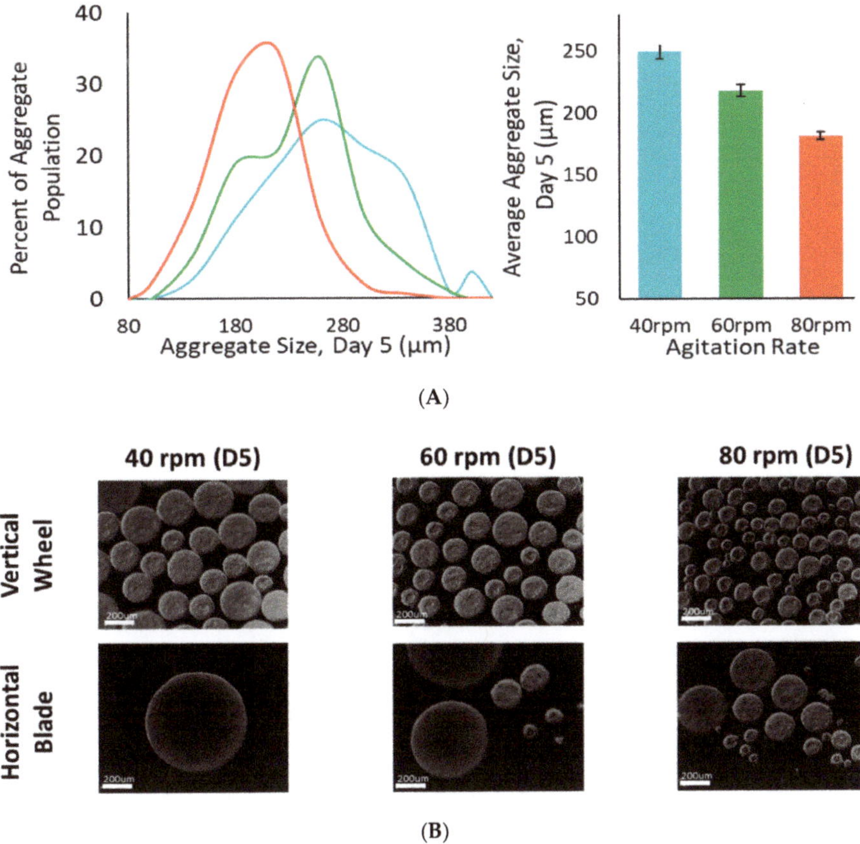

Figure 3. (**A**) Inverse correlation between average iPSC aggregate diameter and agitation rates in Vertical-Wheel bioreactor (0.1 L Scale). (**B**) Comparison of iPSC aggregate diameters and morphology with different agitation rates in Vertical-Wheel bioreactor vs. horizontal-blade spinner (0.1 L scale). Scale bar (200 µm).

As previously mentioned, the uniformity of PSC aggregates is essential not only for the efficiency of cell expansion, but also for directed differentiation, by minimizing the chance that growth and differentiation factors diffuse unevenly through aggregates, and therefore avoiding the undesired heterogeneity and lower quality of target cells. In another set of experiments, iPSC aggregates were successfully differentiated into neural cells, which then formed cerebellar organoids in suspension within small scale Vertical-Wheel bioreactors (Figure 4). After 35 days of a cell expansion and differentiation process, iPSC-derived organoids were efficiently maturated to GABAergic and Glutamatergic neurons in 0.1 L scale Vertical-Wheel bioreactors [23]. Other iPSC aggregates were successfully differentiated into beta cells, cardiomyocytes, or mammary organoids in suspension using different differentiation processes in various scales of Vertical-Wheel bioreactors (data not shown). In addition, studies to differentiate iPSCs into liver organoids in suspension are currently ongoing.

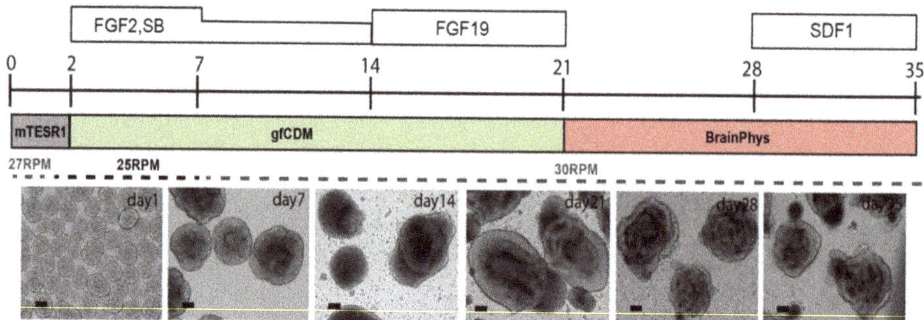

Figure 4. Directed differentiation of human iPSCs into cerebellar organoids in 0.1 L scale Vertical-Wheel bioreactor. Scale bar (100 μm).

While these various experiments were performed at relatively small scales (0.1 to 15 L) in Vertical-Wheel bioreactors, the homogeneous hydrodynamic environment created by the Vertical-Wheel impeller and U-shaped vessel has been modelled by CFD analysis up to 80 L scale and beyond. Therefore, it can be predicted that the formation of uniformly sized PSC aggregates will also be scalable in Vertical-Wheel bioreactors, resulting in efficient cell expansion and differentiation processes even at larger scales.

A PSC expansion process can start with a small cryopreserved vial of PSCs from a working cell bank, inoculating thawed cells into a small-scale bioreactor, and then expanding the cells by serially passaging them into progressively larger bioreactors. Using the scalable bioreactor system with a consistent hydrodynamic environment will improve the efficiency of serial passaging and total cell yield. Studies were performed in order to determine the consistency of PSC expansion in Vertical-Wheel bioreactors, from small-seed culture scale (0.1 L) to potential production scale (80 L). Freshly thawed iPSCs were inoculated into a 0.1 L Vertical-Wheel bioreactor and then four consecutive passages of iPSCs were performed using scale-down model bioreactors of identical size. These studies repeatedly demonstrated that the use of Vertical-Wheel bioreactors led to an average 32-fold expansion of iPSCs during 6–8 days of culture period per passage, achieving a cumulative cell expansion of greater than one million-folds in 28 days (Figure 5). The cells harvested at the end of the serial passage were of high quality, maintaining a normal karyotype, pluripotency, and the ability to form teratomas in vivo. This scale-down model process suggests that commercial scale production of over one trillion high quality iPSCs could feasibly be generated in a Vertical-Wheel bioreactor with 50 L working volume [22].

Various scaling factors can be used to the determine potential scalability of expansion and differentiation processes developed at a small scale in Vertical-Wheel bioreactors. Common parameters that would need to be kept constant during scale-up include Reynolds number, impeller tip speed, power input, maximum shear stress, velocity, and energy dissipation rate. A recent study has suggested that maintaining the volume average energy dissipation rate, determined through computational fluid dynamics simulations, is the best scale-up method for predicting the agitation rates that will sustain a desired average aggregate diameter with narrow size distribution [24]. Therefore, the agitation rates for each progressively larger Vertical-Wheel bioreactor can be calculated that will maintain the same homogenous hydrodynamic environment that was determined during process development in small-scale Vertical-Wheel bioreactors.

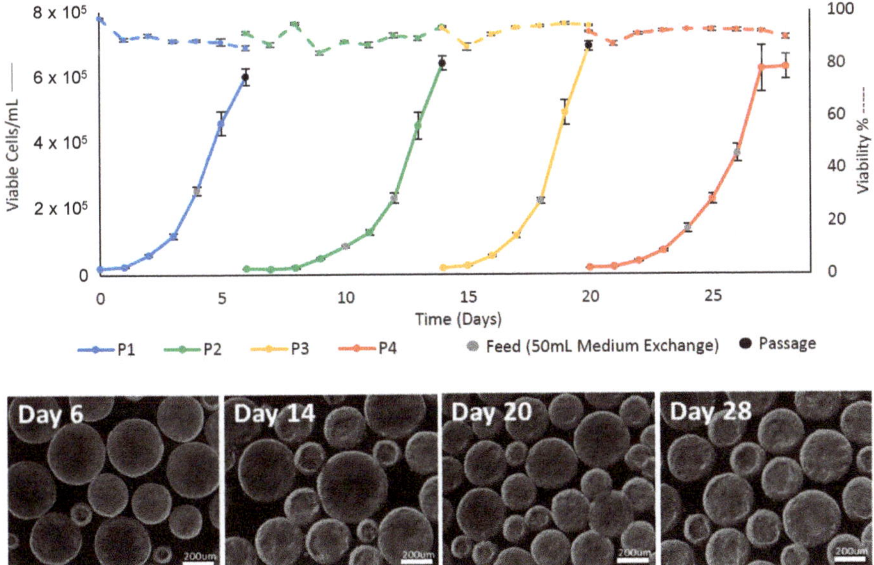

Figure 5. Successful Serial Passaging of iPSCs in 0.1 L Scale Vertical-Wheel Bioreactors.

4. Conclusions

The ultimate goal of providing allogeneic PSC-derived products reliably to vast numbers of patients requires a series of optimized unit operations at various scales to meet target manufacturing lot sizes. Numerous manufacturing processes need to be considered, including seed train, expansion, differentiation, cell harvest, wash, concentration, fill-finish, and cryopreservation. In particular, upstream cell culture processes represent the most challenging bottleneck to achieving robust and economical manufacturing at larger scales. With homogeneous hydrodynamic environments and unparalleled scalability, Vertical-Wheel bioreactors can serve as the innovative manufacturing platforms to enable commercial-scale production of PSC-derived therapeutic cells and to secure the supply of these novel therapeutic cell products to patients.

Author Contributions: B.S.B. and M.S.K. performed the experiments of PSC aggregate expansion, including serial passaging. C.A.V.R., T.P.S., and J.M.S.C. performed the experiments of differentiating PSC aggregates into various organoids. B.L. wrote the manuscript. All authors have given approval of the final version of the manuscript.

Funding: The research in this manuscript received no external funding.

Conflicts of Interest: The authors declare no conflict of interest.

References

1. Takahashi, K.; Yamanaka, S. Induced Puripotent Stem Cells in Medicine and Biology. *Development* **2013**, *140*, 2457–2461. [CrossRef] [PubMed]
2. Shi, Y.; Inoue, H.; Wu, J.C. Induced Pluripotent Stem Cell Technology: A Decade of Progress. *Nat. Rev. Drug Discov.* **2017**, *16*, 115–130. [CrossRef] [PubMed]
3. Simaria, A.S.; Hassan, S.; Varadaraju, H.; Rowley, J.; Warren, K.; Vanek, P.; Farid, S.S. Allogeneic Cell Therapy Bioprocess Economics and Optimization: Single-Use Cell Expansion Technologies. *Biotechnol. Bioeng.* **2014**, *111*, 69–83. [CrossRef] [PubMed]

4. Rodrigues, C.A.V.; Fernandes, T.G.; Diogo, M.M. Stem Cell Cultivation in Bioreactors. *Biotechnol. Adv.* **2011**, *29*, 815–829. [CrossRef] [PubMed]
5. Rodrigues, C.A.V.; Branco, M.; Nogueira, D.E.S.; Silva, T.; Gomes, A.R.; Diogo, M.M.; Cabral, J.M.S. Chapter 2—Bioreactors for Human Pluripotent Stem Cell Expansion and Differentiation. In *Bioreactors for Stem Cell Expansion and Differentiation*; Cabral, J.M.S., Lobato de Silva, Eds.; CRC Press: Boca Raton, FL, USA, 2018.
6. Chen, K.G.; Mallon, B.S.; McKay, R.D.G. Human Pluripotent Stem Cell Culture: Considerations for Maintenance, Expansion, and Therapeutics. *Cell Stem Cell* **2014**, *14*, 13–26. [CrossRef] [PubMed]
7. Nogueira, D.E.S.; Rodrigues, C.A.V.; Carvalho, M.S.; Miranda, C.C.; Hashimura, Y.; Jung, S.; Lee, B.; Cabral, J.M.S. Strategies for The Expansion of Human Induced Pluripotent Stem cells as Aggregates in Single-Use Vertical-Wheel Bioreactors. *J. Biol. Eng.* **2019**, *13*, 74. [CrossRef]
8. Bauwens, C.L.; Peerani, R.; Niebruegge, S.; Woodhouse, K.A.; Kumacheva, E.; Husain, M.; Zandstra, P.W. Control of Human Embryonic Stem Cell Colony and Aggregate Size Heterogeneity Influences Differentiation Trajectories. *Stem Cells* **2008**, *26*, 2300–2310. [CrossRef]
9. Xie, A.W.; Binder, B.Y.K.; Khalili, A.S.; Schmitt, S.K.; Johnson, H.J.; Zacharias, N.A.; Murphy, W.L. Controlled Self-Assembly of Stem Cell Aggregates Instructs Pluripotency and Lineage Bias. *Sci. Rep.* **2017**, *7*, 1–15. [CrossRef]
10. van Winkle, A.P.; Gates, I.D.; Kallos, M.S. Mass transfer limitations in embryoid bodies during human embryonic stem cell differentiation. *Cells Tissues Organs* **2012**, *196*, 34–47. [CrossRef]
11. Wu, J.; Rostami, M.R.; Cadavid Olaya, D.P.; Tzanakakis, E.S. Oxygen Transport and Stem Cell Aggregation in Stirred-Suspension Bioreactor Cultures. *PLos ONE* **2014**, *9*, e102486. [CrossRef]
12. Nogueira, D.E.S.; Rodrigues, C.A.V.; Hashimura, Y.; Jung, S.; Lee, B.; Cabral, J.M.S. Suspension Culture of Human Induced Pluripotent Stem Cells in Single-Use Vertical-Wheel Bioreactors Using Aggregate and Microcarrier Culture Systems. *Methods Mol. Biol.* **2019**, in press.
13. Amit, M.; Margulets, V.; Laevsky, I.; Shariki, K. Suspension Culture of Undifferentiated Human Embryonic Stem Cells and Induced Pluripotent Stem Cells. *Stem Cell Rev.* **2010**, *6*, 248–259. [CrossRef] [PubMed]
14. Hwang, Y.Y.; Chung, B.G.; Ortmann, D.; Hattori, N.; Moeller, H.C.; Khademhosseini, A. Microwell-Mediated Control of Embryoid Body Size Regulates Embryonic Stem Cell Fate via Differential Expression of WNT5a and WNT11. *Proc. Natl. Acad. Sci. USA* **2009**, *106*, 16978–16983. [CrossRef] [PubMed]
15. Miranda, C.; Fernandes, T.G.; Pascoal, J.F.; Haupt, S.; Brüstle, O.; Cabral, J.; Diogo, M.M. Spatial and Temporal Control of Cell Aggregation Efficiently Directs Human Pluripotent Stem Cells Towards Neural Commitment. *Biotechnol. J.* **2015**, *10*, 1612–1624. [CrossRef]
16. Vosough, M.; Omidinia, E.; Kadivar, M.; Shokrgozar, M.A.; Pournasr, B.; Aghdami, N.; Baharvand, H. Generation of Functional Hepatocyte-Like Cells from Human Pluripotent Stem Cells in a Scalable Suspension Culture. *Stem Cells Dev.* **2013**, *22*, 2693–2705. [CrossRef]
17. Kempf, H.; Andree, B.; Zweigerdt, R. Large-Scale Production of Human Pluripotent Stem Cell Derived Cardiomyocytes. *Adv. Drug Deliv. Rev.* **2016**, *96*, 18–30. [CrossRef]
18. Kinney, M.A.; Sargent, C.Y.; McDevitt, T.C. The Multiparametric Effects of Hydrodynamic Environments on Stem Cell Culture. *Tissue Eng. Part B* **2011**, *17*, 249–262. [CrossRef]
19. Mollet, M.; Ma, N.; Zhao, Y.; Brodkey, R.; Taticek, R.; Chalmers, J.J. Bioprocess Equipment: Characterization of Energy Dissipation Rate and its Potential to Damage Cells. *Biotechnol. Prog.* **2004**, *20*, 1437–1448. [CrossRef]
20. Kaiser, S.C.; Löffelholz, C.; Werner, S.; Eibl, D. Chapter 4—CFD for Characterizing Standard and Single-Use Stirred Cell Culture Bioreactors. In *Computational Fluid Dynamics Technologies and Applications*; Minin, O., Ed.; IntechOpen: London, UK, 2011; pp. 97–122.
21. Croughan, M.S.; Giroux, D.; Fang, D.; Lee, B. Chapter 5—Novel Single-Use Bioreactors for Scale-Up of Anchorage-Dependent Cell Manufacturing for Cell Therapies. In *Stem Cell Manufacturing*; Cabral, J.M.S., Lobato de Silva, C., Chase, L.G., Diogo, M.M., Eds.; Elsevier, B.V.: Amsterdam, The Netherlands, 2016; pp. 105–139.
22. Borys, B.S.; So, T.; Colter, J.; Dang, T.; Roberts, E.L.; Revay, T.; Larijani, L.; Krawetz, R.; Lewis, I.; Argiropoulos, B. Optimized Serial Expansion of Human Induced Pluripotent Stem Cells Using Low Density Inoculation to Generate Clinically Relevant Quantities in Vertical-Wheel Bioreactors. *Stem Cells Transl. Med.* **2020**, in press.

23. Silva, T.P.; Fernandes, T.G.; Nogueira, D.E.S.; Rodrigues, C.A.V.; Bekman, E.P.; Hashimura, Y.; Jung, S.; Lee, B.; Carmo-Fonseca, M.; Cabral, J.M.S. Scalable Generation of Mature Cerebellar Organoids from Human Pluripotent Stem Cells and Characterization by Immunostaining. *JoVE* **2020**, in press.
24. Borys, B.S.; Roberts, E.L.; Le, A.; Kallos, M.S. Scale-Up of Embryonic Stem Cell Aggregate Stirred Suspension Bioreactor Culture Enabled by Computational Fluid Dynamics Modeling. *Biochem. Eng. J.* **2018**, *133*, 157–167. [CrossRef]

Article

Electrospinning Live Cells Using Gelatin and Pullulan

Nasim Nosoudi [1,2,*], Anson Jacob Oommen [2], Savannah Stultz [2], Micah Jordan [2], Seba Aldabel [2], Chandra Hohne [2], James Mosser [2], Bailey Archacki [2], Alliah Turner [2] and Paul Turner [2]

1 Department of biomedical engineering, College of Engineering and Computer Sciences (CECS), Marshall University, Weisberg Family Applied Engineering Complex, Huntington, WV 25755, USA

2 Wright State University, Biomedical, Industrial and Human Factors Engineering, 228 Russ Engineering, 3640 Colonel Glenn Hwy, Dayton, OH 45435, USA; oommen.2@wright.edu (A.J.O.); stultz.9@wright.edu (S.S.); jordan.178@wright.edu (M.J.); aldabel.2@wright.edu (S.A.); hohne.4@wright.edu (C.H.); mosser.3@wright.edu (J.M.); archacki.2@wright.edu (B.A.); turner.361@wright.edu (A.T.); turner.425@wright.edu (P.T.)

* Correspondence: nosoudi@marshall.edu; Tel.: +1-304-696-2695

Received: 3 February 2020; Accepted: 20 February 2020; Published: 22 February 2020

Abstract: Electrospinning is a scaffold production method that utilizes electric force to draw a polymer solution into nanometer-sized fibers. By optimizing the polymer and electrospinning parameters, a scaffold is created with the desired thickness, alignment, and pore size. Traditionally, cells and biological constitutes are implanted into the matrix of the three-dimensional scaffold following electrospinning. Our design simultaneously introduces cells into the scaffold during the electrospinning process at 8 kV. In this study, we achieved 90% viability of adipose tissue-derived stem cells through electrospinning.

Keywords: electrospinning; live-cell electrospinning; tissue engineering; cell seeding; high voltage; viability

1. Introduction

Tissue engineering aims to produce synthetic tissues that maintain, restore, or improve native tissue functions [1]. Engineers utilize the formation of both acellular scaffolds and scaffolds that are seeded with cells to accomplish these objectives. Acellular scaffolds are typically used to define a space for new tissues to develop [2]. These scaffolds serve as an extracellular matrix to promote cell adhesion and growth in vivo. Scaffolds with seeded cells have a greater significance because they closely mimic human tissues. It is essential for cell adhesion and migration to occur within these scaffolds. Overall, scaffolds are usually porous and created by various methods, such as electrospinning, phase-separation, freeze-drying, and self-assembly [3]. They enhance the body's ability to heal itself by providing a biodegradable matrix that can enable cells to grow [1].

Electrospinning is a quick and efficient way to produce scaffolds because it allows the scientist to control parameters such as the sizes of nanofibers and nanopores [4]. Other parameters can be constructed as well based on a careful selection of the polymer and an appropriate solvent [5]. During electrospinning, the polymer will dissolve in a volatile solvent and be loaded into a syringe. This liquid is extruded from the needle tip at a constant rate by a syringe pump. In addition, a positive or negative lead will connect to the needle-tip of the syringe while a ground lead is placed on a collector plate [6]. The distance between the syringe-tip and the collector plate will vary depending on the properties of the polymer solution. When the electrostatic force on the polymer solution overcomes the surface tension, a jet of the polymer solution will form and eventually travel towards the collector plate [7]. As the jet flows towards the collector plate, the liquid will accelerate and deposit micro/nanofibers of the polymer on the collector plate [8].

Utilizing current methods, cells are seeded onto the scaffold after it has been formed. Cell seeding can be time-consuming because it requires three steps: creation of the scaffold, differentiation of the cells, and incorporation of the cells into the scaffold. Cell differentiation is already time-consuming and requires additional components, such as growth factors [9]. Another problem arises once these cells are differentiated and seeded: limited ability of cell diffusion into the scaffold [10]. Limited diffusion can produce a nonuniform distribution of cells that causes varied properties and cell densities within different areas of the scaffold. This is potentially detrimental to the longevity of the scaffold both in vitro and in vivo [11]. A possible method to increase cell dispersion in the scaffold is to directly incorporate the cells into the electrospinning process.

There is evidence suggesting that externally applied magnetic fields can affect cell differentiation. It is likely that the generated electric field affects the cell membrane [12]. When the membrane is forced to change shape, it will distort the cytoskeleton of the cell, which attaches the cell membrane to the nucleus. This change in the cytoskeleton will affect the expressed genes and cause the creation of different cell signals, which could induce differentiation [13]. Incorporating stem cells into the electrospinning process will expose them to an electric field that likely induces unique behaviors (i.e., cell differentiation) as previously reported while using lower voltages [14,15]. While this has the potential to be successful, there are many potential hazards to consider.

A concern in this process is that cells could not survive the voltage used in electrospinning. While an electric field could cause unique behaviors, an excessively large electric field could be detrimental to the viability of the cells. Moreover, the field could denature specific protein channels in the membrane, which causes irreparable cell damage [16]. Voltages will be kept as low as possible to prevent this from occurring. Typical electrospinning voltages range from 1 kV to 30 kV [17]. The applied voltage required to create a scaffold will vary depending on the polymer used.

Jayasinghe et al. made use of the coaxial electrospinning from immortalized human brain astrocytoma [18]; a year later, the author described its use from primary porcine vascular smooth muscle cells or rabbit aorta smooth muscle cells, while the protectant polymer was polydimethylsiloxane (PDMS) [19]. Yunmin et al. showed simultaneous bio-electro spraying of human adipose stem cells (ASCs) while electrospinning polyvinyl alcohol (PVA), but the study used two separate needles [20]. Recently, Hoare et al. used hydrazide-functionalized POEGMA (POH) and aldehyde-functionalized POEGMA (POA) along with poly(ethylene oxide) (PEO) to successfully encapsulate NIH 3T3 fibroblasts and electrospin them [21].

This research will test two different polymer combinations: collagen/poly(ethylene oxide) (PEO) and gelatin/pullulan. These polymer combinations will electrospin at around 8 kV of applied voltage. One other constraint for this experiment is the solvent used for dissolving the polymer. In current electrospinning methods, common solvents for collagen include 1,1,1,3,3,3-hexafluoro-2-propanol (HFP), 2,2,2-trifluoroethanol (TFE), or acids (tri-fluoro acetic acid (TFA), acetic acid, hydrochloric acid) [22,23]. These solvents could be toxic if cells are directly incorporated into the polymer–solvent solution. To overcome this restriction, the group will use cell media as the solvent.

Collagen was chosen as the initial polymer because it is the primary constituent of the body's natural extracellular matrix [24]. However, collagen is typically electrospun with an acetic acid solvent, which would likely cause cell death. No studies have attempted to show the success of electrospinning collagen with cell media as the solvent. Therefore, other polymers will also be utilized to determine which one creates the best scaffold. Moreover, collagen may degrade while electrospinning, but successful trials have been reported using different methods, including modifying the collagen surface with methyl methacrylate-*co*-ethyl acrylate [25]. Gelatin is simply denatured collagen; therefore, it can create scaffolds with the same success as collagen [26]. Because previous studies have determined that PEO increases the yield of uniform fibers when electrospun with other polymers, we decided to use it as well [27]. Pullulan and gelatin are commonly used together in hydrogels, and pullulan has shown antioxidant potential [28]. Therefore, the group will electrospin with pullulan, gelatin, or a combination of both. Adipose-derived stem cells (ADSCs) will be used due to their accessibility and potential for creating various terminally differentiated cells, such as osteoblasts, chondrocytes, adipocytes, and neurons [29]. The cells will be directly incorporated into the five polymer solutions prior to electrospinning. To group's knowledge, this is the first time that electrospinning live cells using these natural biocompatible polymers has been reported.

2. Materials and Methods

2.1. Electrospinning Device

Due to the nature of working with living stem cells, it was imperative to maintain sterile conditions throughout the entire spinning process. To maintain a sterile environment, the spinning took place under a sterile biological safety cabinet. The electrospinning device needed to withstand exposure to ultraviolet (UV) light, so it could be sterilized for at least 24 hours under the culture hood as the UV light was usually turned on the night before the experiment. Acrylic, which can handle UV exposure, was determined to be the material that the electrospinning device would be constructed with. Acrylic sheets and cement were used to construct the framework of the electrospinning device along with the necessary spinning supplies, such as a plate, voltage supply, electrical leads, and syringe pump.

2.2. Cell Culturing and Electrospinning

P_2-P_4 of adipose tissue-derived stem cells (hASCs) from Lonza (Walkersville, MD, USA) were used for cell cultures. Cells were plated in T75 culture-treated flasks with approximately 1 million cells per flask, and culture media was changed every 3-4 days for the duration of the culture. Three components make up the cell electrospinning solution: protectant, solvent, and cell pellet. Collagen, Poly(ethylene oxide 10,0000), pullulan, and gelatin powders were used as protectants. Poly(ethylene oxide) (Sigma), pullulan (Hayashibara Laboratories, Okayama, Japan), Type A gelatin from porcine skin (Electron Microscopy Sciences, Hatfield, PA, USA), and extracted collagen from rat tail were dissolved in solvent at concentrations of 2.5 mg/mL, 5 mg/mL, 10 mg/mL, 20 mg/mL, and 30 mg/mL. For PEO, gelatin, and pullulan, the powder was dissolved in serum-free culture media in the previously mentioned concentrations, while collagen acetic acid was used instead of serum-free culture media. The protectants and mesenchymal stem cell medium (solvent) were mixed again at the ratio of 1:1 by volume and placed on a stirring hot plate for 20–30 minutes to warm and mix. The solution was warmed to 40 degrees Celsius in the case of gelatin. The tube temperature reduced to 37 degrees Celsius. Cell pellets (1×10^6) were then added to the protectant solution. Cell electrospinning content was aseptically transferred to a sterile 10 ml syringe, and a sterile 18-gauge syringe needle tip was secured. The collector plate, which is a petri dish, was positioned 7.5 cm from the end of the needle tip. The syringe pump settings were adjusted to produce readings for a plastic 10 ml syringe pump. The pump rate was set to 30 μL/min and reduced at increments of 5 μL/min to determine the optimized pump rate for each cell electrospinning solution. Control cells using the same combination of gelatin, pullulan, gelatin/pullulan, collagen, or PEO were sprayed at the same rate on an empty petri dish without any voltage application. This procedure was done at room temperature, and electrospinning was never performed for more than 15 minutes.

2.3. Viability Test

The viability was investigated by a live/dead assay kit and fluorescence microscopy. Approximately 6 hours after electrospinning, the culture media was aspirated from each well. After incubation with calcein and ethidium (2 μM calcein and 4 μM ethidium in PBS) for 10 minutes at 37 °C, samples were washed with PBS and cells were imaged.

Cytotoxicity Test (Lactate Dehydrogenase (LDH) Activity)

The media was aspirated two days after spinning, and cells were washed with PBS. Lactate dehydrogenase or LDH (Cytotox96 kit, Promega, Madison) was performed on the attached cells according to the manufacture's protocol to look at the cell viability using cell lysate [30].

$$\text{Viability\%} = \frac{Average\ OD\ of\ sample * 100}{Average\ OD\ of\ control}$$

2.4. *Gene Expression by Reverse Transcription-Polymerase Chain Reaction (RT-PCR)*

Seven days after spinning, RNA was isolated according to the manufacturer's instructions for the RNeasy plus mini kit (Qiagen, Germantown, MD, USA), and RT-PCR was performed according to the instruction manual of the One-Step RT-PCR kit (Qiagen, Germantown, MD, USA). The selected pluripotential genes were SOX2 and OCT4.

2.5. *Immunocytochemistry*

Cellular morphology was visualized on Day 2 using fluorescence microscopy. Briefly, samples were fixed with 4% paraformaldehyde (PFA) in PBS (pH 7.4) for 15 min at room temperature (RT). After rinsing with PBS three times, the samples were placed in a permeabilization solution with 0.1% (v/v) Triton X-100 for 10 min and rinsed again with fresh PBS three times. The cells were incubated with Phalloidin 488 and DAPI (Life Technologies, Carlsbad, CA, USA) to visualize the f-actin and nuclei, respectively.

2.6. *Microscopy*

To observe the structure of the scaffold, Fluorescein Isothiocyanate (FITC)-conjugated gelatin was used. Electrospun cells/scaffolds deposited on microscope glass slides were imaged using an Olympus BX51 microscope equipped with an Olympus DP73 camera and CellSens software.

2.7. *CytoViva Microscopy*

To confirm that the cells were embedded within the scaffold, the cells were labeled using a green CMFDA cell tracker dye (Invitrogen, Oregon, Germantown, MD, USA) before electrospinning; they were labeled with DAPI afterward. Samples were imaged using CytoViva's patented enhanced darkfield transmitted light condenser (NA 1.2–1.4) coupled with CytoViva's proprietary Dual Mode Fluorescence (DMF) module. These components were configured on an Olympus BX51 upright microscope using an Olympus100X oil UPL Fluorite objective (NA 0.60–1.30) with adjustable iris objective optimized for darkfield imaging. The light source used was Prior Lumen 200 with a metal halide lamp and variable light attenuation. Optical images were captured using a DAGE-MTI XLMCT cooled CCD camera with a 7.4 µm pixel size.

2.8. *Fourier-Transform Infrared Spectroscopy (FTIR)*

Scaffold compositions were determined by loading the samples onto an attenuated total reflectance (ATR) attachment and using a Thermo Scientific Nicolet iS 50 FTIR (Thermo Fisher, Waltham, MA, USA). Data were plotted in MS Excel (Microsoft, Redmond, WA, USA).

3. Results

A scheme of the electrospinning process is shown in Figure 1A. Electrospinning was only observed at a concentration of 5 mg/mL at 8 kV (Table 1). Cells were detected 6 hours after electrospinning to observe attachment as a sign of viability. Most cells in the collagen scaffold were dead (i.e., stained red). The PEO scaffold had a lot of red cells floating in the petri dish. Gelatin, pullulan, and pullulan/gelatin had good cell viability (i.e., stained fluorescent green), while the number of dead cells (stained red) was minor.

Table 1. Different concentrations that have been tried to electrospin cells.

Polymer	Concentration	Viability from Dead/Live
PEO	2.5 mg/mL, 5 mg/mL, 10 mg/mL, 20 mg/mL, and 30 mg/mL	Not acceptable
Collagen	2.5 mg/mL, 5 mg/mL, 10 mg/mL, 20 mg/mL, and 30 mg/mL	Not acceptable
Gelatin	2.5 mg/mL, 5 mg/mL, 10 mg/mL, 20 mg/mL, and 30 mg/mL	Acceptable
Pullulan	2.5 mg/mL, 5 mg/mL, 10 mg/mL, 20 mg/mL, and 30 mg/mL	Acceptable
Gelatin/ Pullulan	2.5 mg/mL, 5 mg/mL, 10 mg/mL, 20 mg/mL, and 30 mg/mL	Acceptable

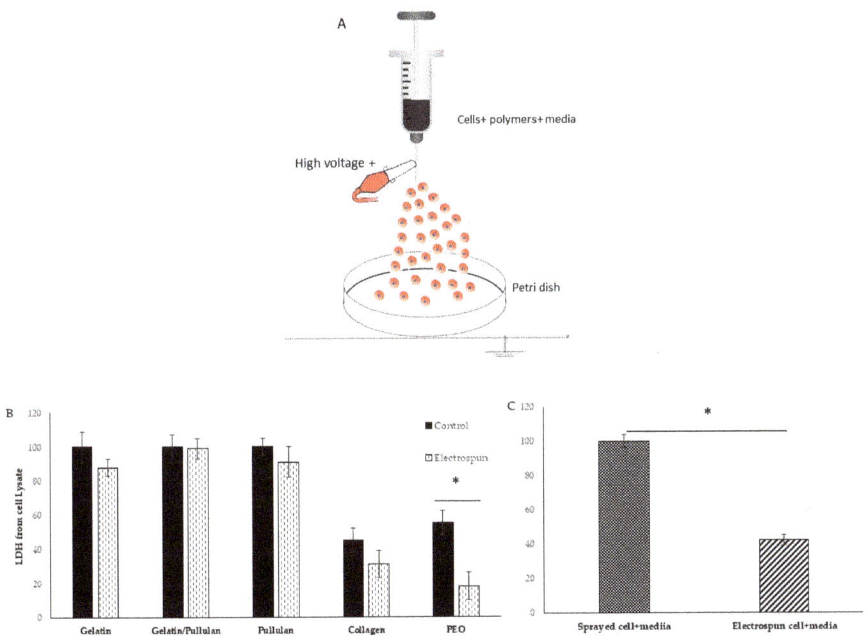

Figure 1. (**A**) Schematic diagram of electrospinning setup for live cells; (**B**) lactate dehydrogenase (LDH) from cell lysate in electrospun and control groups; (**C**) LDH from cell lysate in electrospun (cell+ media) compared to Sprayed cell+ media. Results are normalized to control. * shows p-value < 0.05.

The viability of cells in collagen was very low in both control and electrospun groups. Control cells were sprayed at the same rate on the petri dish without any voltage application. The 0.01% acetic acid that was used to dissolve collagen is probably the reason for low cell viability. PEO was dissolved in cell media and is biocompatible. However, very low cell attachment was observed in both control and electrospun groups, but the group realized this did not result from just the PEO. The PEO viability was 50% in the control group and less than 20% in the electrospun group. There is a significant difference between the control and electrospun group which is caused by the additional effect of electrospinning. In the electrospinning process, the polymer solution is exposed to shear stress, and dead cells in PEO can be the result of non-Newtonian fluids' behaviors and shear stress. When PEO was removed from the formulation, gelatin/cells' viability and attachment were good, and LDH on cell lysate showed 88% cell viability from the electrospun group compared to the control group (Figure 1B,C). However, when the sample was switched to pullulan/gelatin/cells, the group achieved

99% viability compared to the control, but the pullulan/cell scaffold had 91% viability. The *p*-value was not statistically significant among the three groups. To prove the role of the protectants (gelatin and pullulan), cells were electrospun with only culture media, and the cell viability reduced to 40% compared to cells and media that were sprayed with the same rate on the petri dish.

Seven days after electrospinning, Oil Red O, toluidine blue, and Alizarin Red S staining were used to study adipogenic, chondrogenic, and osteogenic differentiations. All cells were negative for Oil Red O, toluidine blue, and Alizarin Red S. Moreover, PCR data showed no significant change in SOX2 and OCT4 after electrospinning (Figure 2), which confirms stemness after and before electrospinning.

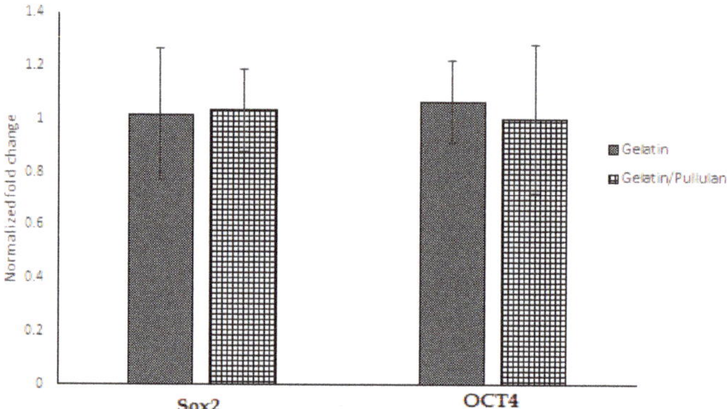

Figure 2. Gene expression in gelatin and gelatin/pullulan electrospun groups normalized to control. Control groups are cells cultured with 5 mg/mL gelatin or 5 mg/mL pullulan/gelatin.

To look at the cell alignment, the group used actin staining 2 days after electrospinning. Cell alignment was random as expected (Figure 3A,B). Images of the scaffold with FITC gelatin showed a porous structure which was later confirmed by CytoViva imaging as well. (Figure 3C).

A highly porous structure was observed after CytoViva imaging. It appears that the cells become embedded in these pores as confirmed by another CytoViva imaging, where cells were pre-stained with a cell tracker and DAPI. Those that house the cells were approximately 10 μm in diameter.

The band observed at 996 cm^{-1} in the pullulan and electrospun samples, which is associated with C-OH bending vibrations at the C-6-position in the case of polysaccharide, indicates the strength of the interchain interactions via hydrogen bonding [31].

The primary hydroxyl groups at the C-6-position were available in the pullulan macromolecule (Figure 4C). However, there were no hydroxyl groups at the C-6-position in gelatin. This band can show the glycosylation between the gelatin and pullulan molecules or the formation of the interchain hydrogen bond in the composite fiber. The amide I (AmI) band at 1630 cm^{-1} in pullulan/gelatin was the strongest among the three and shifted slightly to a higher wavelength, which can be associated with AmI sensitivity to hydrogen bonding at the C=O group [32,33]. Hydrogen bonding plays a significant role in the stabilization of protein secondary structure which can result from the presence of pullulan here [34].

This experiment was run at 8 kV, and the best concentration for Pullulan/gelatin was 5 mg/ml at the ratio of 1:1. Further studies are needed to look at the effect of higher voltage on cell viability and differentiation.

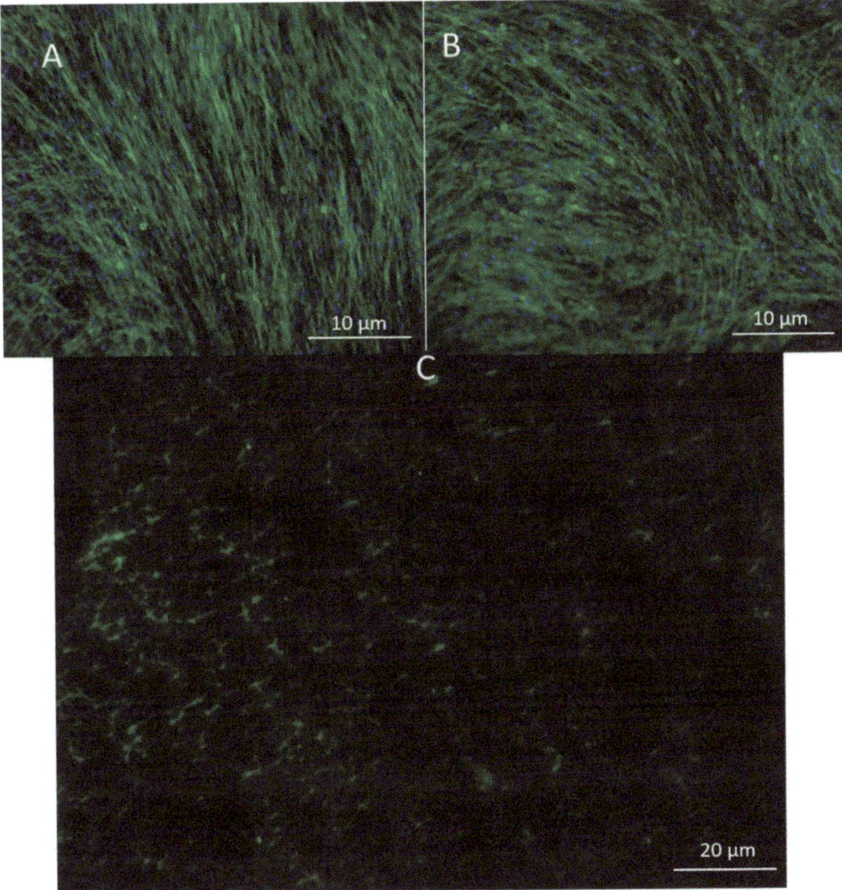

Figure 3. (**A**) Actin staining of adipose-derived stem cells in control and in (**B**) pullulan/gelatin/cells at 10×. Phalloidin 488 (green) labels actin, while DAPI (blue) labels the nucleus. (**C**) Cells were surrounded by FITC (green) conjugated gelatin.

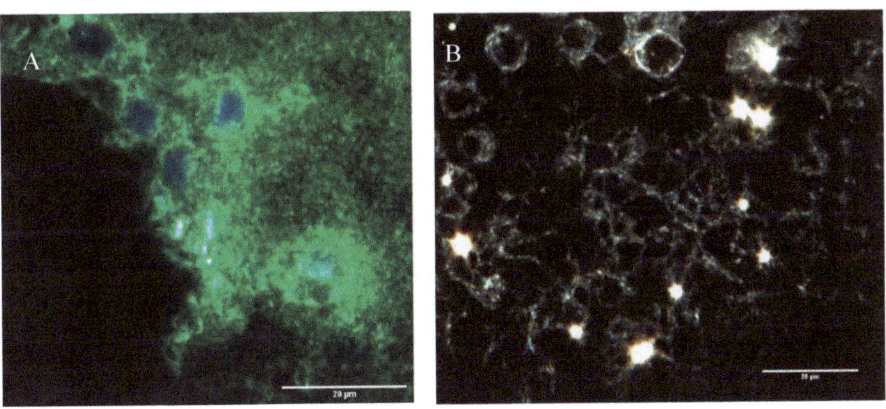

Figure 4. *Cont.*

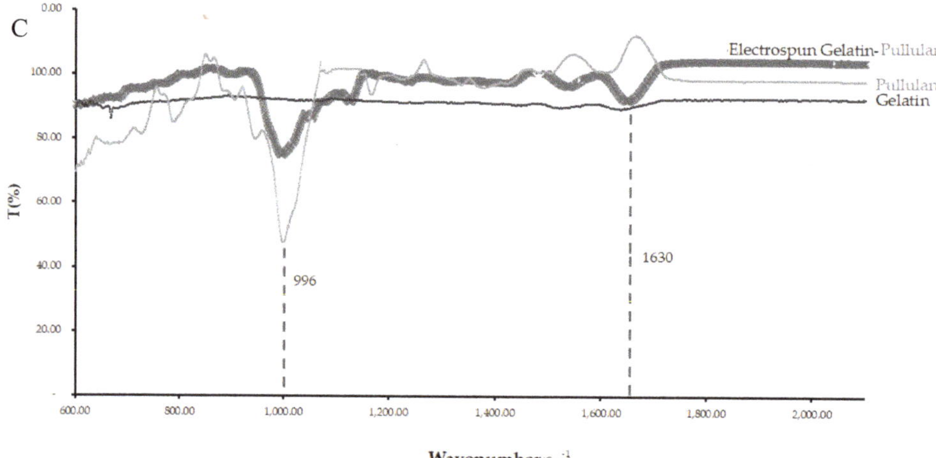

Figure 4. CytoViva and FTIR of the three scaffolds. (**A**) The cells were stained with CellTracker Green CMFDA (Invitrogen) at 2.5 μM for 1 h before electrospinning and were stained for DAPI after electrospinning. (**B**) CytoViva image of the cells and scaffold with no pre-staining. (**C**) FTIR of the pullulan, gelatin and gelatin/pullulan electrospun scaffolds.

4. Discussion

Uniaxial electrospinning with a single needle is a common technology for the fabrication of scaffolds that can provide the initial scaffold for tissue engineering applications. On the other hand, coaxial electrospinning facilitates the incorporation and preservation of bioactive substances, where the shell is often used to protect sensitive substances encapsulated in the core. In this new method, uniaxial electrospinning is incorporated with live cells. Polymers are used to protect the cells, and the cells are encapsulated in the polymers during the electrospinning process. In this study, the group used pullulan, gelatin, collagen, and PEO. Trials with collagen and PEO were unsuccessful as cell viability was not acceptable. Highly hydrated polymers, such as PEO, suppress cellular and molecular adhesions by providing a physical steric barrier [35,36].

This study proved that pullulan and gelatin could protect cells from high voltage damages. Pullulan and gelatin are biocompatible, water-soluble polymers that have been shown to be ineffective at changing phenotype, viability, and cell differentiation. Pullulan can quench reactive oxygen species [37] and be a great scaffold in combination with gelatin. Moreover, pullulan can increase the tensile strength of gelatin, which is very important in tissue engineering [38]. Its structural features, such as the presence of large amounts of hydroxyl groups in the main chain, make it an optimal polymer for creating scaffolds. Studies show that the increase in the pullulan content of a scaffold leads to an increase in viscosity and eventually a decrease in electrical conductivity [39]. The group believes that the composition used in this experiment for making scaffolds acted as a shield for live cells against electrical conductivity, which was shown by the viability studies.

CytoViva images showed a porous structure with cells embedded in it. Moreover, we observed single cells covered by pullulan and gelatin, which was confirmed by fluorescence microscopy. Studies have shown that integrin is an electric field-sensing protein on the cell surface [40]. In addition, gelatin attaches to cells via integrin. Blocking the sensing proteins may be the reason for protecting the cells from high-voltage damage. In general, gelatin can protect the cells by covering the essential structures required for cell function and viability [41]. Gelatin and pullulan are both water-soluble, and this scaffold will dissolve after incubation at 37 °C. These polymers protect the cells during electrospinning but eventually dissolve in the media which can be altered by biocompatible crosslinkers [42].

In this study, no significant difference was found in Sox2 and Oct4 gene expression before and after electrospinning, but it has been shown by numerous studies that low voltage electrical stimulation can affect gene expression of transforming growth factor-β (TGF-β), collagen type-I, alkaline phosphatase (ALP), bone morphogenetic proteins (BMPs), and chondrocyte matrix [43]. More experiments are needed to better understand the effect of 8 kV on stem cells, and the broader gene expression still needs to be studied.

5. Conclusions

The success of this project opens a new field of study within tissue engineering. The discovery that cells can be directly incorporated into the electrospinning process has many potential benefits within the tissue engineering realm. In this paper, electrospinning with an applied voltage of 8 kV was observed at a concentration of 5 mg/mL gelatin/pullulan, but there are many combinations of polymers and cell-types that can still be tested. The application of this design is endless, and many properties of cells may change with variations of voltages and materials. Although this method is limited to water-soluble polymers, using the core–shell technology allows the use of other polymers. By using the core–shell technology, the outer polymer can be loaded with cells and the inner core can be loaded with a polymer that is not water-soluble.

Author Contributions: S.A., J.M., and C.H. performed most of the gelatin studies. P.T., M.J., and A.T. performed most of the gelatin/pullulan studies. S.S., A.J.O., and B.A. helped in microscopy, CytoViva, and preparing the manuscript. N.N. initiated the research and participated in study design and analyses. N.N. edited the final version of this manuscript. All authors have read and agreed to the published version of the manuscript.

Funding: Marshall University start-up fund.

Acknowledgments: I would like to thank Byron Cheatham from Cytoviva and David Ladle from Wright state university for Microscopy. Philippe Georgel, Charles C. Somerville and David Mallory who provided me lab space to finish this work.

Conflicts of Interest: We declare that the authors have no competing interests that might be perceived to influence the results and/or discussion reported in this paper.

References

1. Barnes, C.P.; Sell, S.A.; Boland, E.D.; Simpson, D.G.; Bowlin, G.L. Nanofiber Technology: Designing the Next Generation of Tissue Engineering Scaffolds. *Adv. Drug Deliv. Rev.* **2007**, *59*, 1413–1433. [CrossRef]
2. Glowacki, J.; Mizuno, S. Collagen scaffolds for tissue engineering. *Biopolym. Orig. Res. Biomol.* **2008**, *89*, 338–344. [CrossRef] [PubMed]
3. Chan, B.; Leong, K. Scaffolding in tissue engineering: General approaches and tissue-specific considerations. *Eur. Spine J.* **2008**, *17*, 467–479. [CrossRef] [PubMed]
4. Jun, I.; Han, H.S.; Edwards, J.R.; Jeon, H. Electrospun fibrous scaffolds for tissue engineering: Viewpoints on architecture and fabrication. *Int. J. Mol. Sci.* **2018**, *19*, 745. [CrossRef] [PubMed]
5. Frenot, A.; Chronakis, I.S. Polymer nanofibers assembled by electrospinning. *Curr. Opin. Colloid Interface Sci.* **2003**, *8*, 64–75. [CrossRef]
6. Doshi, J.; Reneker, D.H. Electrospinning process and applications of electrospun fibers. *J. Electrost.* **1995**, *35*, 151–160. [CrossRef]
7. Lu, T.; Li, Y.; Chen, T. Techniques for fabrication and construction of three-dimensional scaffolds for tissue engineering. *Int. J. Nanomed.* **2013**, *8*, 337. [CrossRef]
8. Olson, J.L.; Atala, A.; Yoo, J.J. Tissue engineering: Current strategies and future directions. *Chonnam Med. J.* **2011**, *47*, 1–13. [CrossRef]
9. Gimble, J.M.; Guilak, F. Adipose-derived adult stem cells: Isolation, characterization, and differentiation potential. *Cytotherapy* **2003**, *5*, 362–369. [CrossRef]
10. Solchaga, L.A.; Tognana, E.; Penick, K.; Baskaran, H.; Goldberg, V.M.; Caplan, A.I.; Welter, J.F. A rapid seeding technique for the assembly of large cell/scaffold composite constructs. *Tissue Eng.* **2006**, *12*, 1851–1863. [CrossRef]

11. Nosoudi, N.; Holman, D.; Karamched, S.; Lei, Y.; Rodriguez-Dévora, J. Engineered Extracellular Matrix: Current Accomplishments and Future Trends. *Int. J. Biomed. Eng. Sci.* **2014**, *1*, 1–15.
12. Sauer, H.; Rahimi, G.; Hescheler, J.; Wartenberg, M. Effects of electrical fields on cardiomyocyte differentiation of embryonic stem cells. *J. Cell. Biochem.* **1999**, *75*, 710–723. [CrossRef]
13. Schwartz, L.; Moreira, J.d.; Jolicoeur, M. Physical forces modulate cell differentiation and proliferation processes. *J. Cell. Mol. Med.* **2018**, *22*, 738–745. [CrossRef] [PubMed]
14. Zhao, H.; Steiger, A.; Nohner, M.; Ye, H. Specific intensity direct current (DC) electric field improves neural stem cell migration and enhances differentiation towards βIII-tubulin+ neurons. *PLoS ONE* **2015**, *10*, e0129625. [CrossRef]
15. Jaatinen, L. The Effect of an Applied Electric Current on Cell Proliferation, Viability, Morphology, Adhesion, and Stem Cell Differentiation . Ph.D. Thesis, Tampere University of Technology, Tampere, Finland, 2017.
16. Chen, W. Electroconformational denaturation of membrane proteins. *Ann. N. Y. Acad. Sci.* **2006**, *1066*, 92–105. [CrossRef]
17. Li, D.; Xia, Y. Electrospinning of nanofibers: Reinventing the wheel? *Adv. Mater.* **2004**, *16*, 1151–1170. [CrossRef]
18. Townsend-Nicholson, A.; Jayasinghe, S.N. Cell electrospinning: A unique biotechnique for encapsulating living organisms for generating active biological microthreads/scaffolds. *Biomacromolecules* **2006**, *7*, 3364–3369. [CrossRef]
19. Jayasinghe, S.N.; Irvine, S.; McEwan, J.R. Cell electrospinning highly concentrated cellular suspensions containing primary living organisms into cell-bearing threads and scaffolds. *Future Med.* **2007**. [CrossRef]
20. Yunmin, M.; Yuanyuan, L.; Haiping, C.; Qingxi, H. Application and analysis of biological electrospray in tissue engineering. *Open Biomed. Eng. J.* **2015**, *9*, 133. [CrossRef]
21. Xu, F.; Dodd, M.; Sheardown, H.; Hoare, T. Single-Step Reactive Electrospinning of Cell-Loaded Nanofibrous Scaffolds as Ready-to-Use Tissue Patches. *Biomacromolecules* **2018**, *19*, 4182–4192. [CrossRef]
22. Bhattarai, D.P.; Aguilar, L.E.; Park, C.H.; Kim, C.S. A review on properties of natural and synthetic based electrospun fibrous materials for bone tissue engineering. *Membranes* **2018**, *8*, 62. [CrossRef] [PubMed]
23. Bazrafshan, Z.; Stylios, G.K. Spinnability of collagen as a biomimetic material: A review. *Int. J. Biol. Macromol.* **2019**, *129*, 693–705. [CrossRef] [PubMed]
24. Babitha, S.; Rachita, L.; Karthikeyan, K.; Shoba, E.; Janani, I.; Poornima, B.; Sai, K.P. Electrospun protein nanofibers in healthcare: A review. *Int. J. Pharm.* **2017**, *523*, 52–90. [CrossRef] [PubMed]
25. Bazrafshan, Z.; Stylios, G.K. A novel approach to enhance the spinnability of collagen fibers by graft polymerization. *Mater. Sci. Eng. C* **2019**, *94*, 108–116. [CrossRef]
26. Li, G.Y.; Fukunaga, S.; Takenouchi, K.; Nakamura, F. Comparative study of the physiological properties of collagen, gelatin and collagen hydrolysate as cosmetic materials. *Int. J. Cosmet. Sci.* **2005**, *27*, 101–106. [CrossRef]
27. Huang, L.; Nagapudi, K.; PApkarian, R.; Chaikof, E.L. Engineered collagen–PEO nanofibers and fabrics. *J. Biomater. Sci.* **2001**, *12*, 979–993. [CrossRef]
28. Nicholas, M.N.; Jeschke, M.G.; Amini-Nik, S. Cellularized bilayer pullulan-gelatin hydrogel for skin regeneration. *Tissue Eng. Part A* **2016**, *22*, 754–764. [CrossRef]
29. Frese, L.; Dijkman, P.E.; Hoerstrup, S.P. Adipose tissue-derived stem cells in regenerative medicine. *Transfus. Med. Hemotherapy* **2016**, *43*, 268–274. [CrossRef]
30. Mobini, S.; Hoyer, B.; Solati-Hashjin, M.; Lode, A.; Nosoudi, N.; Samadikuchaksaraei, A.; Gelinsky, M. Fabrication and characterization of regenerated silk scaffolds reinforced with natural silk fibers for bone tissue engineering. *J. Biomed. Mater. Res. Part A* **2013**, *101*, 2392–2404. [CrossRef]
31. Li, R.; Tomasula, P.; De Sousa AM, M.; Liu, S.C.; Tunick, M.; Liu, K.; Liu, L. Electrospinning pullulan fibers from salt solutions. *Polymers* **2017**, *9*, 32. [CrossRef]
32. Zhang, C.; Gao, D.; Ma, Y.; Zhao, X. Effect of gelatin addition on properties of pullulan films. *J. Food Sci.* **2013**, *78*, C805–C810. [CrossRef] [PubMed]
33. Myshakina, N.S.; Ahmed, Z.; Asher, S.A. Dependence of amide vibrations on hydrogen bonding. *J. Phys. Chem. B* **2008**, *112*, 11873–11877. [CrossRef] [PubMed]
34. Liguori, A.; Uranga, J.; Panzavolta, S.; Guerrero, P.; de la Caba, K.; Focarete, M.L. Electrospinning of Fish Gelatin Solution Containing Citric Acid: An Environmentally Friendly Approach to Prepare Crosslinked Gelatin Fibers. *Materials* **2019**, *12*, 2808. [CrossRef] [PubMed]

35. Vacheethasanee, K.; Wang, S.; Qiu, Y.; Marchant, R.E. Poly (ethylene oxide) surfactant polymers. *J. Biomater. Sci. Polym. Ed.* **2004**, *15*, 95–110. [CrossRef]
36. Bulman, S.E.; Coleman, C.M.; Murphy, J.M.; Medcalf, N.; Ryan, A.E.; Barry, F. Pullulan: A new cytoadhesive for cell-mediated cartilage repair. *Stem Cell Res. Ther.* **2015**, *6*, 34. [CrossRef]
37. Xu, Q.; He, C.; Xiao, C.; Chen, X. Reactive oxygen species (ROS) responsive polymers for biomedical applications. *Macromol. Biosci.* **2016**, *16*, 635–646. [CrossRef]
38. Lim, L.-T.; Mendes, A.C.; Chronakis, I.S. Electrospinning and electrospraying technologies for food applications. *Adv. Food Nutr. Res.* **2019**, *88*, 167–234.
39. Topuz, F.; Uyar, T. Influence of Hydrogen-Bonding Additives on Electrospinning of Cyclodextrin Nanofibers. *ACS Omega* **2018**, *3*, 18311–18322. [CrossRef]
40. Tsai, C.H.; Lin, B.J.; Chao, P.H.G. $\alpha 2\beta 1$ integrin and RhoA mediates electric field-induced ligament fibroblast migration directionality. *J. Orthop. Res.* **2013**, *31*, 322–327. [CrossRef]
41. Davidenko, N.; Schuster, C.F.; Bax, D.V.; Farndale, R.W.; Hamaia, S.; Best, S.M.; Cameron, R.E. Evaluation of cell binding to collagen and gelatin: A study of the effect of *2D* and *3D* architecture and surface chemistry. *J. Mater. Sci. Mater. Med.* **2016**, *27*, 148. [CrossRef]
42. Sinha, A.; Nosoudi, N.; Vyavahare, N. Elasto-regenerative properties of polyphenols. *Biochem. Biophys. Res. Commun.* **2014**, *444*, 205–211. [CrossRef] [PubMed]
43. Meng, S.; Rouabhia, M.; Zhang, Z.; De, D.; De, F.; Laval, U. Electrical stimulation in tissue regeneration. *Appl. Biomed. Eng.* **2011**, 37–62.

Review

Culture Time Needed to Scale up Infrapatellar Fat Pad Derived Stem Cells for Cartilage Regeneration: A Systematic Review

Sam L. Francis [1,2,3,*], Angela Yao [2] and Peter F. M. Choong [1,2,3]

1 Department of Surgery, The University of Melbourne, Melbourne, VIC 3065, Australia; pfchoong@gmail.com
2 Department of Orthopaedics, St Vincent's Hospital, Melbourne, VIC 3056, Australia; angela.yao.inbox@gmail.com
3 Biofab 3D, Aikenhead Centre for Medical Discovery, Melbourne, VIC 3065, Australia
* Correspondence: sam2703@gmail.com; Tel.: +61-466-640-801

Received: 12 June 2020; Accepted: 2 July 2020; Published: 4 July 2020

Abstract: Adipose tissue is a rich source of stem cells, which are reported to represent 2% of the stromal vascular fraction (SVF). The infrapatellar fat pad (IFP) is a unique source of tissue, from which human adipose-derived stem cells (hADSCs) have been shown to harbour high chondrogenic potential. This review aims to calculate, based on the literature, the culture time needed before an average knee articular cartilage defect can be treated using stem cells obtained from arthroscopically or openly harvested IFP. Firstly, a systematic literature review was performed to search for studies that included the number of stem cells isolated from the IFP. Subsequent analysis was conducted to identify the amount of IFP tissue harvestable, stem cell count and the overall yield based on the harvesting method. We then determined the minimum time required before treating an average-sized knee articular cartilage defect with IFP-derived hADSCs by using our newly devised equation. The amount of fat tissue, the SVF cell count and the stem cell yield are all lower in arthroscopically harvested IFP tissue compared to that collected using arthrotomy. As an extrapolation, we show that an average knee defect can be treated in 20 or 17 days using arthroscopically or openly harvested IFP-derived hADSCs, respectively. In summary, the systematic review conducted in this study reveals that there is a higher amount of fat tissue, SVF cell count and overall yield (cells/volume or cells/gram) associated with open (arthrotomy) compared to arthroscopic IFP harvest. In addition to these review findings, we demonstrate that our novel framework can give an indication about the culture time needed to scale up IFP-derived stem cells for the treatment of articular cartilage defects based on harvesting method.

Keywords: stromal vascular fraction; stem cell; adipose-derived stem cell; infrapatellar fat pad; knee; arthroscopy; arthrotomy; tissue engineering

---

1. Introduction

Human adipose-derived mesenchymal stem cells (hADSCs) [1–5] are a significant player in tissue engineering, particularly for forming chondrogenic tissue [6,7]. The two primary adipose sources harvested are the subcutaneous fat tissue and the infrapatellar fat pad (IFP). The IFP, which is also known as Hoffa's fat pad, is a distinct bulk of fat that sits behind the patellar tendon extending into the anterior knee joint [8]. The tissue is functionally structural adipose with little or no metabolic response; furthermore, its contribution to knee function is unclear if any [8]. Stem cells isolated from the IFP have been shown to produce higher chondrogenic potential in comparison to those sourced from subcutaneous fat [9] and possess significant multi-differentiation potential [10]. IFP-derived hADSCs have been used for intra-articular injections into osteoarthritic human knees [11] and for engineered tissue implants for in vivo cartilage repair [12,13].

In the literature, in vivo animal models of osteochondral repair using IFP-derived hADSCs paired with tissue engineering techniques utilise the following treatment process: (1) harvest and isolation of hADSCs, (2) culture expansion of hADSCs, (3) reimplantation of hADSCs combined with a scaffold [12]. A significant concern in these studies has been the use of excessive cell-expansion timeframes (step 2 of the treatment process), which can be performed for many months and passages. Long-term cell expansion in a laboratory is associated with an increased risk of tumorigenic transformation, contamination and exposure to animal-based serum products [14].

When envisioning future clinical translation of regenerative therapies in different individuals, an efficient expansion timeframe needs to be identified. Another key advantage of using an efficient duration for reimplantation is to identify the patient waiting time between surgical procedures (harvest and reimplantation).

To identify the minimum time required for cellular expansion before treatment, three critical pieces of data are needed; (1) the number of hADSCs required for therapy, (2) the IFP-derived hADSCs doubling time, and (3) the number of hADSCs that can be isolated from the IFP.

The cell concentration in healthy cartilage and the defect volume in question dictate how many stem cells are required for therapy. The average number of cells per mL of articular cartilage was described as 1.0×10^7 by Hunziker et al. (2002) [15]. The average knee defect volume in humans is approximately 550 μL (mm^3) as shown in a hallmark study reporting the findings of diagnostic arthroscopies [16]. Therefore, in an average-sized defect, 5.5×10^6 cells are required for regenerative therapy. The doubling time of IFP-derived hADSCs has been described as roughly 5 days by Garcia et al. (2016) [17]. The remaining piece of data needed is the number of stem cells that can be isolated from the IFP, which has been poorly studied, with no extensive case studies evaluating the cellular yield. When the IFP is digested, the initial cellular population retrieved is a stromal vascular fraction (SVF), which contains many different types of cells. The proportion of SVF cells that are hADSCs has been described in the literature as roughly 2% [18,19], and to specifically isolate these cells, selective plastic adherence and cellular expansion are required as per the International Society for Cellular Therapy [20,21].

A key factor influencing the cellular yield is the amount of fat tissue harvestable, which is limited when using the IFP. The IFP is generally removed along with surrounding connective and fibrous tissue, meaning the overall fat content is less than the amount of total tissue collected. Traditionally the two methods to harvest IFP tissue are keyhole surgery (arthroscopy) or formal opening of the knee joint (open arthrotomy) during arthroplasty. The amount of IPF tissue that can be harvested by either technique is different, given one procedure is minimally invasive and the other is open.

Hence, the overall aim of this systematic review is to identify the culture time needed to scale up IFP-derived stem cells for cartilage regeneration. The data gap within the literature is the evaluation of how many cells there are in the IFP. By identifying this information, we can couple it with the number of cells needed for therapy and the doubling rate (described above) to ascertain the minimum timeframe required for cellular expansion before reimplantation and therapy. This is calculated using a modified derivation of the cellular doubling time equation [17] and is detailed in the methods section.

The specific outcomes that will be assessed are; the amount of fat tissue retrievable, the cellular yield and finally, the number of SVF cells obtainable. These outcomes will be evaluated by the method of harvest used (arthroscopic or open arthrotomy).

Finally, using the sourced data from the systematic review, the minimum time required for cellular expansion prior to treating an average-sized knee defect is determined using a newly derived equation by the authors.

2. Materials and Methods

2.1. Study Design and Literature Search

A systematic literature review based on the Preferred Reporting Items for Systematic Reviews and Meta-Analyses (PRISMA) guidelines [22] was conducted on the 15th of October, 2019.

Two independent investigators (SF, AY) searched electronic databases (PubMed, Cochrane Central Register of Controlled Trials (CENTRAL) and Embase) with English language restrictions.

The following keywords were used: "infrapatellar fat pad" AND "mesenchymal stem cell" OR "adipose-derived stem cell". Duplicates were removed and reference lists were also searched manually for additional studies. Our search combined the following search term groups;

1. Infrapatellar fat pad
2. Stem cell OR mesenchymal stem cell OR adipose-derived mesenchymal stem cell

Using the Boolean operator OR within the search group and then combining the two groups with the operator AND we formulated our search design.

2.2. Study Selection and Eligibility Criteria

The search was restricted to human studies written in English published within the last 20 years (1998 onwards). Search results from each database were then imported into EndNote bibliographic management software (Thomson Reuters, Endnote version X7, Ottawa, ON, Canada) and duplicates were then identified. Level one screening (SF and AY) was performed by looking at article titles and abstract records and screened based on carefully selected inclusion and exclusion criteria, as shown in Table 1.

Table 1. Inclusion and exclusion criteria.

Inclusion Criteria	Exclusion Criteria
Human participants ≥ 18 years old	Conference abstracts and non-original papers (commentaries, reviews, letters or editorials)
Studies written in English	Animal studies
Clinical studies, multi-centre studies, observational studies, clinical trials, randomised control trials, meta-analyses and systematic reviews	Studies conducted in children, adolescents or pregnant women
Measurement of MSCs derived from IFP only	Non-IFP-derived MSCs or studies lacking specific data relating to IFP or MSC numbers
Explicit measurement of the quantitative value of MSCs found per participant	Off-topic studies

The remaining pool of articles underwent level 2 screening (same exclusion parameters as level 1). The full text of these articles was screened by two independent reviewers (SF and AY) and relevant articles selected.

Differences in opinion were resolved by consensus. Additional hand searching of reference lists of included full-text articles were identified and added for inclusion if appropriate.

2.3. Data Extraction

Two independent reviewers (SF, AY) obtained relevant data and assessed the accuracy of the screened articles. In the event that both authors could not reach an agreement, a third author (PC) decided. The following information was extracted from each study: first author's name, year of publication, country, study design, participant eligibility, patient demographics, study sample size, harvest procedure, surgical indication, the quantity of IFP tissue and cells yielded per study.

Contact with corresponding authors was also attempted when deemed necessary to verify the accuracy of the data and obtain further data for the analysis.

2.4. Quality Control and Assessment

Quality control procedures for articles included level 1 (titles/abstracts) and level 2 (full text) screening for eligibility according to inclusion and exclusion criteria. Level 2 screening, as mentioned prior, was performed independently by two researchers. Articles for which there was any uncertainty about inclusion were discussed with a third researcher. Data were extracted from full-text versions of articles.

2.5. Minimum Time Required for Regenerative Therapy Using IFP-Derived hADSCs

This is calculated using the data sourced from the systematic review and the following equation, which is a derivation from the cell population doubling time equation [17]. The number of days in culture ($t2-t1$) from the original equation corresponds to the minimum time (days) used in this formula.

$$\text{Minimum time (days)} = [\text{DT} \times \ln(n2/n1)]/[\ln(2)] \quad (1)$$

Theorem 1. *Where DT refers to the cell doubling time, n2 is the total number of hADSCs required for therapy and n1 is the number of hADSCs initially isolated. The number of hADSCs is calculated as 2% of the SVF population, based on literature described in the introduction.*

3. Results

3.1. Search Results

Using the search strategy described above, 156 articles were obtained, of which 72 were screened at level 1 after review of title and abstracts. Of the remaining 84 articles, after reviewing the full text, a further 72 articles were screened out at level 2. References were checked in the remaining 12 articles, and no new articles were added to the final list. Reasons for article exclusions are detailed in Figure 1.

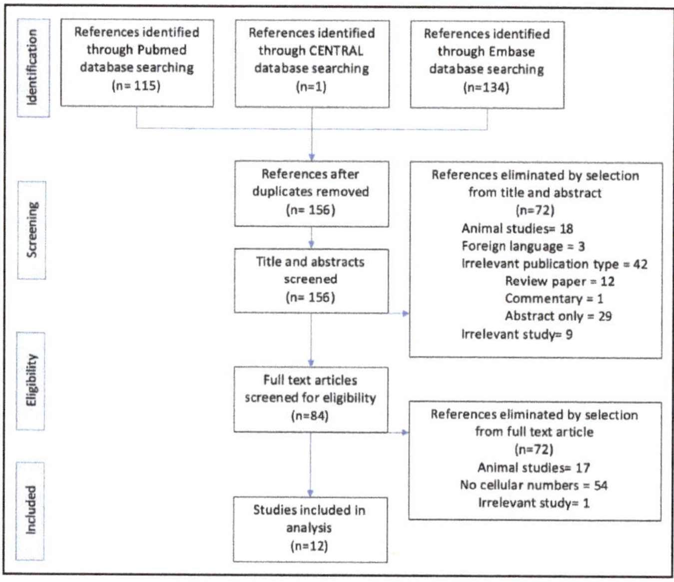

Figure 1. Article screening process, conducted using the PRISMA guidelines, Liberati et al., BMJ, 2009 [22].

3.2. Included Studies

All 12 included articles were controlled single-institution studies (Table 2). Of the 12 isolated studies, 11 were retrospective [23–33] and one was a prospective cohort study [34]. There was a total of 228 patient IFPs assessed in the 12 studies. The patient ages ranged between 17 and 89. From the six studies indicating gender [23–25,27,31,34], 77 (62%) patients were female and 48 (38%) were male.

The amount of IFP sourced was described by volume in three (25%) studies [23,25,28], by weight in six (50%) studies [26,30–34] and not described in three (25%) studies [24,27,29]. The range of IFP harvested by volume was described between 2.5 and 25 mL, while the range, when described by weight, was between 5 and 26.3 g. The number of cells (SVF) at isolation was reported in 10 (92%) studies [23–32,34] and was described as either the total number of cells isolated, cells/g or cells/fat pad. The remaining study [33] did not report the number of cells at isolation; instead, the number of cells obtained after two passages of tissue culture was described. The yield of cells will be presented more in detail below when specifically assessed based on harvesting method.

3.3. Study Outcomes

3.3.1. Arthroscopic Based Harvest of IFP

There were 3 (25%) studies [25,27,34] that investigated arthroscopically harvested IFP. In total, 53 patient IFPs were assessed in these three studies. Patient ages ranged between 17 and 69. All three papers indicated gender, with 27 (51%) patients being female and 26 (49%) being male. The amount of IFP sourced was described by volume in one (33%) study [25], by weight in 1 (33%) study [34] and not specified in the remaining (33%) study [27]. One study presented the cellular count as a mesenchymal stem cell (MSC) count; however, this conflicted with their methodology and was likely to represent the SVF count [34].

The study [25] describing the volume of tissue harvested assessed 3 mL of IFP, from which an average of 7.0×10^5 (6.0–8.0) SVF cells were retrieved. The yield by volume in this study was, on average, 2.3×10^5 cells/mL of IFP tissue. The study describing the weight of IFP tissue harvested averaged 9.4 g (6.9–11.2), with an average SVF count of 1.89×10^6 ($1.2–2.3 \times 10^6$). The yield by weight in these studies was, on average, 2.01×10^5 cells/g of IFP.

The overall yield in all these studies was calculated to be 1.2×10^6 cells per IFP ($0.93–2.3 \times 10^6$), irrespective of whether volume or weight was used to describe the harvested amount.

3.3.2. Open Arthrotomy Based Harvest of IFP

There were 10 (77%) studies [23–26,29–31] that investigated the open harvest of IFP using arthrotomy. A total of 175 patient IFPs were assessed, with ages ranging between 37 and 89. Gender was indicated in four studies [23–25,31]; 40 (65%) patients were female and 22 (35%) were male. The amount of IFP sourced was described by volume in three (30%) studies [23,25,28], by weight in five (50%) studies [26,30–33] and not described in two (20%) studies. One study did not describe the SVF yield obtained during cellular isolation.

The volume of tissue harvested was on average 4.4 mL of IFP (2.5–25); in these studies, the number of SVF cells isolated on average was 7.9×10^5 cells (0.1–120). The yield by volume in these studies on average was 1.8×10^5 cells/mL of IFP tissue. The weight of IFP tissue harvested was on average 15.2 g (5.0–26.3); in these studies, the number of SVF cells isolated on average was 4.5×10^6 cells (3.94–5.5). The yield by weight in these studies was, on average, 3.0×10^5 cells/g of IFP.

The overall yield in all these studies was calculated to be 3.0×10^6 cells per IFP ($2.8–4.0 \times 10^6$), irrespective of whether volume or weight was used to describe the harvested amount.

Table 2. Overview of each study included in the review. TKA—Total knee arthroplasty, OA—Osteoarthritis, SVF—Stromal vascular fraction, IFP—Infrapatellar fat pad. Data is represented as the mean +/− standard deviation (SD).

Study	Country	Patient Demographics	Harvest Procedure (Surgical Indication)	IFP Tissue Yield	Cellular Data/Yield
Mantripragada et al. *2019* [23]	USA	n = 28 Mean age: 63.1 (37–82) F = 16, M = 12	TKA (Idiopathic OA)	2.5 mL of IFP	SVF = 4320 cells per mL of IFP
Lopez-Ruiz et al. *2018* [24]	Spain	n = 8 Mean age: 66.9 (57–74) F = 4, M = 4	TKA (Idiopathic OA)	Not described	SVF = 1.1×10^5 cells per gram of IFP ($0.9–1.3 \times 10^5$)
Bravo et al. *2019* [25]	Spain	n = 21 ≥ Arthroscopic ACL repair Median age: 32 (21–50) F = 4, M = 17 n = 21 ≥ TKA Median age: 74 (57–84) F = 15, M = 6	Arthroscopic repair (ACL rupture) Or TKA (OA)	≥3 mL of adipose extracted from IFP ≥3 mL of adipose extracted from IFP	$6.0 \times 10^5–8.0 \times 10^5$ cells
Wang et al. *2017* [26]	China	n = 4	TKA (No stated indication)	10 g of IFP	Primary culture (SVF) ≥ Approx. 5×10^5/g of IFP
Dragoo et al. *2017* [27]	USA	n = 7 Median age: 35.14 (17–52) F = 6, M = 1	Arthroscopic repair (ACL rupture with no evidence of OA)	Not described	SVF = 4.86×10^5 cells $0.19–0.33 \times 10^5$ cells per gram of IFP
Munoz-Criado et al. *2017* [29]	Spain	n = 24 Ages: 50–80	TKA (Idiopathic OA)	Not described	SVF = $7.8 \times 10^5 \pm 2.8 \times 10^5$ cells
Neri et al. *2017* [30]	Italy	n = 11 Mean age: 69.4 ± 6.5	TKA (Idiopathic OA)	5–15 g IFP obtained	SVF = Average of 7.0×10^5 cells per gram of IFP
Tangchitphisut et al. *2016* [31]	Thailand	n = 5 Mean age: 65.8 (53–77) F = 5, M = 0	TKA (Idiopathic OA)	Average IFP weight 12.12 ± 2.57 g (8.48–14.75)	SVF: $3.94 \times 10^6 \pm 3.73 \times 10^5$ cells
Koh et al. *2012* [34]	South Korea	n = 25 Mean age: 54.2 ± 9.3 (34–69) F = 17, M = 8	Arthroscopic harvest (secondary OA)	Average IFP weight 9.4 g (6.9–11.2)	SVF (described as MSC count in the paper) = 1.89×10^6 ($1.2–2.3 \times 10^6$)
Jurgens et al. *2009* [32]	Netherlands	n = 53 Median age: 72 (43–89)	TKA (Idiopathic OA)	Average IFP weight 15.1 ± 5.8 g (8.7–26.3)	SVF: $4.0 \times 10^6 \pm 4.45 \times 10^5$ cells
Dragoo et al. *2003* [28]	USA	n = 5 Mean age: 74 (53 to 86)	TKA (no indication stated)	Average IFP volume 20.6 mL (15–25)	SVF mean yield of 5.5×10^6 extracted cells per IFP (2.0×10^6 to 1.2×10^7)
Wickham et al. *2003* [33]	USA	n = 16 Mean age: 68 ± 11.1 (49–82)	TKA (no indication stated)	Average IFP weight 21.5 ± 8.8 g	Not described

4. Discussion

Sourced papers from the systematic review originated from Europe, Asia and America, representing a diverse pool of data; furthermore, a good spread between gender and age groups was obtained. The difference in surgical procedures amongst different regions is an important consideration, although for IFP harvest, arthroscopic and arthrotomy related techniques are generally the same worldwide. Primary osteoarthritis and ligament damage were the indications for treatment in these studies and given that IFP is opportunistically harvested, the lack of literature/data on the IFP of healthy individuals remains a key limitation in this field.

The volume of tissue harvested by arthroscopy was, on average, 3 mL compared to 4.4 mL of IFP using open arthrotomy. The weight of tissue harvested was on average, 9.3 g compared to 15.2 g of IFP tissue using arthroscopy and arthrotomy, respectively. Although safe and efficiently performed, the restricted access and exposure associated with arthroscopic surgery appear to significantly limit the amount of tissue harvestable as compared to arthrotomy, where the whole IFP can be removed easily.

The overall cellular yield was also lower in tissue harvested from arthroscopy compared to arthrotomy, irrespective of whether the amount was described by volume or weight. The overall yield of SVF cells was, on average, 1.2×10^6 and 3.0×10^6 cells per IFP when using arthroscopy and an open arthrotomy, respectively. Arthroscopically harvested tissue contains a fluidic component that is specific to the technique, as opposed to harvest via open arthrotomy, in which the entire IFP is retrieved en mass without the mixing or introduction of any fluid. This dilution does not affect the total number of cells retrieved; however, it can lead to a lower overall yield per unit of tissue. Another consideration as to why the yield is lower arthroscopically is whether the roughness involved with arthroscopic drainage through a tube may lead to cellular damage and loss, therefore, reducing the overall live cell count and yield.

Analysing all the IFP samples, on average 1.2×10^6 and 3.0×10^6 SVF cells were isolated using arthroscopic and open harvest respectively. Therefore, using the formula described in the methods, the earliest time point in which an average-sized knee defect can be treated using hADSCs sourced from arthroscopic or open IFP harvest was calculated to be 40 (35–41 days) and 33 days (31–33 days), respectively (calculation shown in Supplementary Section S1). If both IFPs (one in each human knee) are utilised, treatment can be performed in 20 (17–21 days) and 17 (16–17 days) days respectively. A limitation to the use of this calculation is the high variability seen in SVF counts, the use of an assumed doubling time and assumed hADSCs percentage within the SVF. In culmination these differences can add up to make a large spread of results, to control this, in our calculation, we used the standard deviation range of the SVF yields to present a range of days that it could take to scale cells up. Ultimately, this equation can be used as a guideline and the use of more defined input data allows for a more precise end calculation.

Although it takes longer culturing time to scale up arthroscopically harvested IFP-derived hADSCs to treat articular cartilage defects, this approach is the most likely to be adopted when envisioning future clinical translation of regenerative therapies for osteochondral repair. This is due to the minimally invasive nature of the technique, which is associated with less surgical risk compared to an open arthrotomy and enables the development of a same-day surgical repair option.

Using the same calculation described in this review, the culture time required to scale IFP-derived stem cells prior to cartilage repair can be determined for any defect volume. Therefore, this work represents a framework that has never been described in the literature. The next step is to further validate this framework by evaluating the cells obtained after specific expansion times with respect to phenotype and function. This can be performed by utilizing an in vitro study to prove that cells expanded for specific timeframes display the hADSCs phenotype and can then be successfully driven into chondrogenesis. If this can be achieved in vitro, a subsequent in vivo model could be performed to ultimately prove chondrogenic regeneration using these adapted time calculations in the native joint environment.

5. Conclusions

This systematic review shows that there is a higher amount of fat tissue, SVF cell count and overall yield (cells/volume or cells/gram) associated with open (arthrotomy) compared to arthroscopic IFP harvest. As an extrapolation, it takes an average of 20 or 17 days to scale up arthroscopically or openly harvested IFP-derived hADSCs, respectively, to reach a sufficient amount of cells enabling cartilage defect repair.

Supplementary Materials: The following are available online at http://www.mdpi.com/2306-5354/7/3/69/s1, Section S1: calculation of the minimum time required for regenerative therapy using IFP-derived hADSCs.

Author Contributions: Conceptualization, S.L.F. and P.F.M.C.; methodology, S.L.F. and A.Y.; software, S.L.F.; validation, S.L.F., A.Y. and P.F.M.C.; formal analysis, S.L.F. and A.Y.; investigation, S.L.F. and A.Y.; resources, S.L.F.; data curation, S.L.F. and A.Y.; writing—original draft preparation, S.L.F. and A.Y.; writing—review and editing, S.L.F., A.Y., P.F.M.C.; visualization, S.L.F. and A.Y.; supervision, P.F.M.C.; project administration, S.L.F. All authors have read and agreed to the published version of the manuscript.

Funding: This research received no external funding.

Acknowledgments: The authors would like to acknowledge the support of funding from the Royal Australian College of Surgeons (RACS), The Australian Research Council Industrial Transformation Training Centre (ARC-ITTC) in Additive Biomanufacturing, the Medical Technologies and Pharmaceuticals (MTP Connect) BioMedTech Horizons program, the St Vincent's research endowment fund and the AVANT doctors in training program.

Conflicts of Interest: Authors declare no conflict of interest.

Abbreviations

SVF	Stromal vascular fraction
IFP	Infrapatellar fat pad
MSCs	Mesenchymal stem cells
hADSCs	Human adipose-derived mesenchymal stem cell
CENTRAL	Cochrane Central Register of Controlled Trials
PRISMA	Preferred Reporting Items for Systematic Reviews and Meta-Analyses

References

1. Bunnell, B.A.; Flaat, M.; Gagliardi, C.; Patel, B.; Ripoll, C. Adipose-Derived stem cells: Isolation, expansion and differentiation. *Methods* **2008**, *45*, 115–120. [CrossRef] [PubMed]
2. Zuk, P.A.; Zhu, M.; Mizuno, H.; Huang, J.; Futrell, J.W.; Katz, A.J.; Benhaim, P.; Lorenz, H.P.; Hedrick, M.H. Multilineage Cells from Human Adipose Tissue: Implications for Cell-Based Therapies. *Tissue Eng.* **2001**, *7*, 211–228. [CrossRef] [PubMed]
3. Becker, A.; Mc Culloch, C.E.; Till, J.E. Cytological demonstration of the clonal nature of spleen colonies derived from transplanted mouse marrow cells. *Nature* **1963**, *197*, 452–454. [CrossRef] [PubMed]
4. Friedenstein, A.J.; Deriglasova, U.F.; Kulagina, N.N.; Panasuk, A.F.; Rudakowa, S.F.; Luriá, E.A.; Ruadkow, I.A. Precursors for fibroblasts in different populations of hematopoietic cells as detected by the in vitro colony assay method. *Exp. Hematol.* **1974**, *2*, 83–92. [PubMed]
5. Pittenger, M.; Mackay, A.; Beck, S.; Jaiswal, R.; Douglas, R.; Mosca, J.; Moorman, M.; Simonetti, D.; Craig, S.; Marshak, D. Multilineage Potential of Adult Human Mesenchymal Stem Cells. *Science* **1999**, *284*, 143–147. [CrossRef] [PubMed]
6. Rada, T.; Reis, R.L.; Gomes, M.E. Adipose Tissue-Derived Stem Cells and Their Application in Bone and Cartilage Tissue Engineering. *Tissue Eng. Part B Rev.* **2009**, *15*, 113–125. [CrossRef]
7. Veronesi, F.; Maglio, M.; Tschon, M.; Aldini, N.N.; Fini, M. Adipose-Derived mesenchymal stem cells for cartilage tissue engineering: State-of-The-Art in in vivo studies. *J. Biomed. Mater. Res. Part A* **2013**, *102*, 2448–2466. [CrossRef]
8. Ioan-Facsinay, A.; Kloppenburg, M. An emerging player in knee osteoarthritis: The infrapatellar fat pad. *Arthritis Res.* **2013**, *15*, 225. [CrossRef]

9. Stanco, D.; De Girolamo, L.; Sansone, V.; Moretti, M. Donor-Matched mesenchymal stem cells from knee infrapatellar and subcutaneous adipose tissue of osteoarthritic donors display differential chondrogenic and osteogenic commitment. *Eur. Cells Mater.* **2014**, *27*, 298–311. [CrossRef]
10. Sheng, G. The developmental basis of mesenchymal stem/stromal cells (MSCs). *BMC Dev. Biol.* **2015**, *15*, 44. [CrossRef]
11. Spasovski, D.; Spasovski, V.; Bascarevic, Z.; Stojiljkovic, M.; Vreca, M.; Andelkovic, M.; Pavlović, S. Intra-articular injection of autologous adipose-derived mesenchymal stem cells in the treatment of knee osteoarthritis. *J. Gene Med.* **2018**, *20*, e3002. [CrossRef]
12. Di Bella, C.; Duchi, S.; O'Connell, C.D.; Blanchard, R.; Augustine, C.; Yue, Z.; Thompson, F.; Richards, C.; Beirne, S.; Onofrillo, C.; et al. In situ handheld three-dimensional bioprinting for cartilage regeneration. *J. Tissue Eng. Regen. Med.* **2017**, *12*, 611–621. [CrossRef] [PubMed]
13. Vahedi, P.; Jarolmasjed, S.; Shafaei, H.; Roshangar, L.; Rad, J.S.; Ahmadian, E. In vivo articular cartilage regeneration through infrapatellar adipose tissue derived stem cell in nanofiber polycaprolactone scaffold. *Tissue Cell* **2019**, *57*, 49–56. [CrossRef] [PubMed]
14. Halme, D.G.; Kessler, D.A. FDA Regulation of Stem-Cell–Based Therapies. *N. Engl. J. Med.* **2006**, *355*, 1730–1735. [CrossRef] [PubMed]
15. Hunziker, E. Articular cartilage repair: Basic science and clinical progress. A review of the current status and prospects. *Osteoarthr. Cartil.* **2002**, *10*, 432–463. [CrossRef]
16. Hjelle, K.; Solheim, E.; Strand, T.; Muri, R.; Brittberg, M. Articular cartilage defects in 1000 knee arthroscopies. *Arthroscopy* **2002**, *18*, 730–734. [CrossRef]
17. Garcia, J.; Wright, K.; Roberts, S.; Kuiper, J.H.; Mangham, C.; Richardson, J.; Mennan, C. Characterisation of synovial fluid and infrapatellar fat pad derived mesenchymal stromal cells: The influence of tissue source and inflammatory stimulus. *Sci. Rep.* **2016**, *6*, 24295. [CrossRef]
18. Kokai, L.E.; Traktuev, D.O.; Zhang, L.; Merfeld-Clauss, S.; Dibernardo, G.; Lu, H.; Marra, K.G.; Donnenberg, A.; Meyer, E.M.; Fodor, P.B.; et al. Adipose Stem Cell Function Maintained with Age: An Intra-Subject Study of Long-Term Cryopreserved Cells. *Aesthet. Surg. J.* **2016**, *37*. [CrossRef]
19. Tsekouras, A.; Mantas, D.; Tsilimigras, D.I.; Moris, D.; Kontos, M.; Zografos, C.G. Comparison of the Viability and Yield of Adipose-Derived Stem Cells (ASCs) from Different Donor Areas. *In Vivo* **2017**, *31*, 1229–1234. [CrossRef]
20. Dominici, M.; Le Blanc, K.; Mueller, I.; Slaper-Cortenbach, I.; Marini, F.; Krause, D.; Deans, R.; Keating, A.; Prockop, D.; Horwitz, E. Minimal criteria for defining multipotent mesenchymal stromal cells. The International Society for Cellular Therapy position statement. *Cytotherapy* **2006**, *8*, 315–317. [CrossRef]
21. Krampera, M.; Galipeau, J.; Shi, Y.; Tarte, K.; Sensebe, L. MSC Committee of the International Society for Cellular Therapy (ISCT). Immunological characterization of multipotent mesenchymal stromal cells—The International Society for Cellular Therapy (ISCT) working proposal. *Cytotherapy* **2013**, *15*, 8. [CrossRef] [PubMed]
22. Liberati, A.; Altman, U.G.; Tetzlaff, J.; Mulrow, C.; Gøtzsche, P.C.; Ioannidis, J.P.A.; Clarke, M.; Devereaux, P.J.; Kleijnen, J.; Moher, D. The PRISMA statement for reporting systematic reviews and meta-analyses of studies that evaluate healthcare interventions: Explanation and elaboration. *BMJ* **2009**, *339*, b2700. [CrossRef] [PubMed]
23. Mantripragada, V.P.; Piuzzi, N.S.; Bova, W.A.; Boehm, C.; Obuchowski, N.A.; Lefebvre, V.; Midura, R.J.; Muschler, G. Donor-Matched comparison of chondrogenic progenitors resident in human infrapatellar fat pad, synovium, and periosteum—Implications for cartilage repair. *Connect. Tissue Res.* **2019**, *60*, 597–610. [CrossRef] [PubMed]
24. Lopez-Ruiz, E.; Jimenez, G.; Kwiatkowski, W.; Montanez, E.; Arrebola, F.; Carrillo, E.; Choe, S.; Marchal, J.A.; Perán, M. Impact of TGF-beta family-related growth factors on chondrogenic differentiation of adipose-derived stem cells isolated from lipoaspirates and infrapatellar fat pads of osteoarthritic patients. *Eur. Cell Mater.* **2018**, *35*, 209–224. [CrossRef]
25. Bravo, B.; Guisasola, M.C.; Vaquero, J.; Tirado, I.; Gortazar, A.R.; Forriol, F. Gene expression, protein profiling, and chemotactic activity of infrapatellar fat pad mesenchymal stem cells in pathologies of the knee joint. *J. Cell. Physiol.* **2019**, *234*, 18917–18927. [CrossRef]

26. Wang, B.; Liu, W.; Xing, D.; Li, R.; Lv, C.; Li, Y.; Yan, X.; Ke, Y.; Xu, Y.; Du, Y.; et al. Injectable nanohydroxyapatite-chitosan-gelatin micro-scaffolds induce regeneration of knee subchondral bone lesions. *Sci. Rep.* **2017**, *7*, 16709. [CrossRef]
27. Dragoo, J.; Chang, W. Arthroscopic Harvest of Adipose-Derived Mesenchymal Stem Cells from the Infrapatellar Fat Pad. *Am. J. Sports Med.* **2017**, *45*, 3119–3127. [CrossRef]
28. Dragoo, J.L.; Samimi, B.; Zhu, M.; Hame, S.L.; Thomas, B.J.; Lieberman, J.R.; Hedrick, M.H.; Benhaim, P. Tissue-engineered cartilage and bone using stem cells from human infrapatellar fat pads. *J. Bone Jt. Surg. Br. Vol.* **2003**, *85*, 740–747. [CrossRef]
29. Muñoz-Criado, I.; Meseguer-Ripolles, J.; Mellado-López, M.; Alastrue-Agudo, A.; Griffeth, R.J.; Forteza-Vila, J.; Cugat, R.; García, M.; Moreno-Manzano, V. Human Suprapatellar Fat Pad-Derived Mesenchymal Stem Cells Induce Chondrogenesis and Cartilage Repair in a Model of Severe Osteoarthritis. *Stem Cells Int.* **2017**, *2017*, 1–12. [CrossRef]
30. Neri, S.; Guidotti, S.; Lilli, N.L.; Cattini, L.; Mariani, E. Infrapatellar fat pad-derived mesenchymal stromal cells from osteoarthritis patients: In vitro genetic stability and replicative senescence. *J. Orthop. Res.* **2016**, *35*, 1029–1037. [CrossRef]
31. Tangchitphisut, P.; Srikaew, N.; Numhom, S.; Tangprasittipap, A.; Woratanarat, P.; Wongsak, S.; Kijkunasathian, C.; Hongeng, S.; Murray, I.R.; Tawonsawatruk, T. Infrapatellar Fat Pad: An Alternative Source of Adipose-Derived Mesenchymal Stem Cells. *Arthritis* **2016**, *2016*, 1–10. [CrossRef] [PubMed]
32. Jurgens, W.J.; Van Dijk, A.; Doulabi, B.Z.; Niessen, F.B.; Ritt, M.J.; Van Milligen, F.J.; Helder, M.N. Freshly isolated stromal cells from the infrapatellar fat pad are suitable for a one-step surgical procedure to regenerate cartilage tissue. *Cytotherapy* **2009**, *11*, 1052–1064. [CrossRef] [PubMed]
33. Wickham, M.Q.; Erickson, G.R.; Gimble, J.M.; Vail, T.P.; Guilak, F. Multipotent Stromal Cells Derived from the Infrapatellar Fat Pad of the Knee. *Clin. Orthop. Relat. Res.* **2003**, *412*, 196–212. [CrossRef] [PubMed]
34. Koh, Y.-G.; Choi, Y.-J. Infrapatellar fat pad-derived mesenchymal stem cell therapy for knee osteoarthritis. *Knee* **2012**, *19*, 902–907. [CrossRef] [PubMed]

Review

Production of Human Pluripotent Stem Cell-Derived Hepatic Cell Lineages and Liver Organoids: Current Status and Potential Applications

João P. Cotovio and Tiago G. Fernandes *

iBB-Institute for Bioengineering and Biosciences and Department of Bioengineering, Instituto Superior Técnico, Universidade de Lisboa, Av. Rovisco Pais, 1049-001 Lisbon, Portugal; joaocotovio@tecnico.ulisboa.pt
* Correspondence: tfernandes@tecnico.ulisboa.pt

Received: 14 March 2020; Accepted: 7 April 2020; Published: 9 April 2020

Abstract: Liver disease is one of the leading causes of death worldwide, leading to the death of approximately 2 million people per year. Current therapies include orthotopic liver transplantation, however, donor organ shortage remains a great challenge. In addition, the development of novel therapeutics has been limited due to the lack of in vitro models that mimic in vivo liver physiology. Accordingly, hepatic cell lineages derived from human pluripotent stem cells (hPSCs) represent a promising cell source for liver cell therapy, disease modelling, and drug discovery. Moreover, the development of new culture systems bringing together the multiple liver-specific hepatic cell types triggered the development of hPSC-derived liver organoids. Therefore, these human liver-based platforms hold great potential for clinical applications. In this review, the production of the different hepatic cell lineages from hPSCs, including hepatocytes, as well as the emerging strategies to generate hPSC-derived liver organoids will be assessed, while current biomedical applications will be highlighted.

Keywords: human pluripotent stem cells; hepatic cell lineages; hepatocyte differentiation; non-parenchymal liver cells; liver organoids; disease modeling; drug screening

---

1. Introduction

From all the internal organs that constitute the human body, the liver is the largest one, and its endocrine and exocrine properties make it also the largest gland. As an endocrine gland, the liver is responsible for the secretion of several hormones, while the bile constitutes the major exocrine secretion. The liver is therefore a central organ in our body that is responsible for homeostasis throughout the human lifespan, performing a complex array of functions [1,2]. Such functions include glycogen storage, drug detoxification, control of metabolism, regulation of cholesterol synthesis and transport, urea metabolism, immunological activity, and secretion of plasma proteins like albumin [1]. The liver is such an essential player in homeostasis, that liver disease due to genetic or environmental factors, such as hepatitis, fibrosis, cirrhosis, and hepatocellular carcinoma, often results in morbidity and mortality [3]. Actually, liver disease is one of the leading causes of death worldwide and it is estimated that approximately 2 million people die per year, representing 3.5% of global deaths. From the 2 million, 1.16 million deaths are caused by cirrhosis (11th cause of death worldwide) and 0.79 million deaths are caused by hepatocellular carcinoma (16th cause of death worldwide) [4]. In addition to mortality rates, liver disease is estimated to have an impact on over 600 million people around the world [5]. Besides that, for end-stage liver failure and other disorders, orthotopic liver transplantation is the only possible solution, making the liver the second most common solid organ transplantation. Still, transplantation needs are poorly met [4].

Unfortunately, the discovery of novel therapeutics for liver disease is still a major challenge, as in vitro modeling of the in vivo physiological functions of the liver is still not accurate. In vitro models are traditionally based on hepatocyte cultures since they are the major parenchymal cell type, accounting for 60% of the total cells in the organ (80% of the volume), mediating almost all liver functions and being its functional metabolic unit [2]. The gold standard source of hepatocytes for scientific investigation has been freshly isolated primary human hepatocytes (PHHs). However, when in culture these cells lose their ability to proliferate and most of their functions are impaired. Additionally, they have limited supply and present batch-to-batch variability, which greatly limit their potential for clinical applications [6,7]. Other hepatocyte sources usually rely on immortalized or cancer cell lines, as well as fetal liver progenitors or adult liver stem cells. Nevertheless, these sources also present major drawbacks, like the fact that their metabolic enzymes do not entirely resemble the ones in adult hepatocytes, in addition to poor cell survival, proliferation, and availability [6,7].

Accordingly, the generation of hepatocytes derived from human pluripotent stem cells (hPSCs) represents a promising cell source to transform our understanding of liver disease and to change the way it is treated [6]. However, their remarkable potential can only be translated into practice if the capability to direct the differentiation of hPSCs into the different hepatic lineages can be achieved. In fact, considerable progress has been made in the past few years in the development of defined protocols for hepatocyte differentiation, which will be further addressed in detail in this review. Hepatocytes may be the largest cellular component of the liver, but its complexity goes far beyond that. Along with hepatocytes, other cell types account for 40% of the cells in the liver, including cholangiocytes (biliary epithelial cells) and other non-parenchymal cells, like liver sinusoidal endothelial cells (LSECs), Kupffer cells (resident liver macrophages), pit cells (natural killer cells), and hepatic stellate cells (HSCs) [1]. To prove this complexity and heterogeneity within the human liver, recent studies were able to create a human liver cell atlas [8]. However, the challenge is now to determine how these cells arrange themselves to form the three-dimensional (3D) architecture that is so crucial for hepatic function. The growing knowledge related to hPSC differentiation triggered the development of new 3D culture technologies like organoids [9]. Consequently, the first steps for human liver organoid production are now being made. In this review, the production of the different hepatic cell lineages from hPSCs, as well as the emerging strategies to generate hPSC-derived liver organoids will be assessed, highlighting their great potential for biomedical applications in regenerative medicine, disease modelling, drug discovery, and hepatotoxicity.

2. Hepatogenesis: Origin and Fates of Hepatic Cells

Within the early embryo, as cells develop, they become progressively restricted in their developmental potency. When the pluripotent state is reached, a PSC has the capacity to produce all cell types from the tissues that constitute the human body [10]. However, much of the knowledge in this area is based on mouse developmental processes. During embryogenesis, more precisely during the process of gastrulation, there is the formation of a transient structure in the region of the epiblast, the primitive streak (PS). The developmental path of the liver has its origin prior to, or shortly after, the beginning of gastrulation. Throughout this process, uncommitted cells migrate through the PS and undergo an epithelial-to-mesenchymal transition (EMT), ending in the generation of the three germ layers: ectoderm, mesoderm, and endoderm [11,12]. Thus, the formation of endoderm, the germ layer that eventually gives rise to the liver, results from the intercalation of the definite endoderm that migrated through the PS and minimally by the pre-existing visceral endoderm [13,14]. Molecularly, the TGFβ family member Nodal is responsible for mammalian endoderm specification [15], with high levels of Nodal promoting endoderm and low levels promoting mesoderm specification [16]. This signaling gradient activates the expression of key transcription factors like EOMES (in high levels) and T (in low levels), leading to the cooperation between EOMES with NODAL-SMAD2/3 signaling to induce the expression of endodermal markers such as CER1, FOXA2, and SOX17 [17].

After gastrulation, endoderm undergoes a series of morphogenetic movements resulting in the formation of the gut tube that, in mammals, is patterned along the anterior–posterior axis into three regions, comprising the foregut, midgut, and hindgut [18]. This patterning results from the action of Wnts, bone morphogenetic proteins (BMPs), and fibroblast growth factors (FGFs), that in a gradient from low levels in the anterior region to high levels in the posterior region, results in an anterior foregut, posterior foregut, and midgut-hindgut fate, respectively [19]. The posterior foregut endoderm contains progenitor cells that can give rise to pancreas, liver, and gallbladder. However, for hepatic specification, the convergence of the cardiac mesoderm [20] and septum transversum mesenchyme (STM) [21] is required. FGFs, like FGF1 and FGF2, from the cardiac mesoderm [22,23], as well as BMPs, like BMP4 and BMP2, from the STM are key players in this process [21]. Additionally, endothelial cells also play a relevant role in hepatic specification [24]. Thus, hepatic endoderm thickens to form a liver diverticulum initiating a budding process into the STM. Soon, hepatic endoderm transit from a columnar to a pseudostratified epithelium constituted by hepatoblasts that proliferate and migrate through the STM to form the liver bud [25]. The transcription factor HHEX controls the transition from columnar to pseudostratified epithelium [25], whereas TBX3 and PROX1 promote hepatoblast proliferation and migration [26,27]. HNF4a is equally essential for further development of the liver bud structure [28]. Hepatoblasts are bipotent, having the potential to differentiate into either hepatocytes or cholangiocytes in a process regulated by transforming growth factor-beta (TGFβ), Notch, Wnt, BMP, and FGF [19,29]. Cells committed to a hepatocyte fate progressively mature, and together with cholangiocytes and non-parenchymal cell types, give rise to the final structure of the adult liver (for extend review on liver development see [1,30]).

3. Differentiation of Hepatic Cell Lineages from Human Pluripotent Stem Cells

Since the isolation of embryonic stem cells (ESCs) from the blastocyst [31–33], and later with the recapitulation of pluripotency in vitro by reprogramming somatic cells into induced pluripotent stem cells (iPSCs) [34–36], numerous protocols have been described for the direct differentiation of PSCs into many cell populations across the three germ layers, including neurons [37], cardiomyocytes [38], and also hepatic cells (Table 1). Most of the research in differentiating hepatic cell types from hPSCs has been focused on hepatocytes, but differentiation strategies that originate other enriched populations of liver cells have also been developed. The design of such protocols should bring a new level of complexity to the study of liver development and medical research.

The majority of the differentiation protocols already published are at their core a recapitulation of the natural developmental processes that occur during embryogenesis. Thus, the differentiation of PSCs into a certain cell population can be achieved through the addition of extrinsic signals that guide the cells into a particular fate while repressing alternative ones. These extrinsic signals are usually presented as a combination of signaling pathway agonists and antagonists, added to the culture medium in a stepwise fashion in specific concentrations, sequence, and time [39]. To attain this, it is necessary to understand not only the signaling pathway kinetics, but also the developmental roadmap of the different cell types of the liver, and studying how different cell populations are segregated during fate transitions [40].

3.1. Hepatocytes

Since hepatocytes constitute the major cell type in the liver, there has been great interest in differentiating PSCs into this type of cell. The first reports date back to 1996 and relied on the spontaneous differentiation of mouse ESCs using embryoid bodies (EBs) [41]. These EBs spontaneously recapitulate early steps of embryogenesis in an uncontrolled fashion, ending in a highly variable structure with differentiated cells of all three germ layers, where hepatocytes can be present. Only in the following decade the first direct differentiation of hPSCs was reported using hESCs [42] and latter using hiPSCs [43]. Since then, based on developmental signaling pathways and morphogens, many protocols have used similar growth factors and small molecules to generate hepatocytes in adherent culture conditions, leading to outstanding improvements in efficiency and functionality (Table 1).

As previously described, Nodal/Activin, members of the TGF-β superfamily of signaling molecules, are critical for definitive endoderm induction in mammals, and the same is true for in vitro differentiation of PSCs. Kubo and colleagues were the first to provide evidence that high concentrations of Activin A result in endoderm induction [44]. Since then, this growth factor has been widely used in the first steps of hepatocyte differentiation, sometimes associated with other signaling molecules that also promote endoderm development, like BMPs, FGFs, or WNT3A, among others. Afterwards, to recapitulate hepatic specification, BMPs and FGFs have been used to give rise to hepatic endoderm, FGF2, FGF4, BMP4, and BMP2 being the most commonly used. Finally, to trigger hepatoblast proliferation and subsequent hepatocyte differentiation/maturation, hepatocyte growth factor (HGF), Oncostatin M (OSM), and Dexamethasone (DEX) are among the most frequent choices. These differentiation strategies result in the production of what can be called hepatocyte-like cells (HLCs), since current protocols still do not generate hepatocytes with fully mature phenotype when compared to adult hepatocytes. To simplify, in this review we will only use the term "hepatocyte". Nevertheless, several approaches have been used for in vitro maturation of hPSC-derived hepatocytes, including growth factors, small molecules, transcription factors, and microRNAs [45]. Examples of that are the already mentioned HGF and OSM. It is established that HGF can promote hepatoblast proliferation, migration, and survival, and in the presence of DEX it can upregulate several mature hepatocyte markers [46]. On the other hand, it has also been proven that fetal hepatocytes in the presence of OSM acquire a similar morphology to mature hepatocytes, also contributing to hepatic functionality [46,47].

The majority of the differentiation methods to generate hepatocytes can then be divided into three main stages: endoderm induction, hepatic specification, and hepatocyte differentiation/maturation. However, these three-step approaches may not precisely mimic in vivo liver development, generating impure populations. Trying to overcome this challenge, recent studies attempt to recapitulate hepatocyte development through a sequence of six consecutive lineage choices [40]. On the other hand, there are groups trying to overcome the lack of definition and reproducibility between protocols, relying for that purpose on methods solely driven by small molecules, following a growth-factor-free strategy [48]. Additionally, some studies have been focused on the generation of scalable protocols for large-scale production of hPSC-derived hepatocytes [49–51].

3.2. Cholangiocytes

Apart from hepatocytes, cholangiocytes have a key role in the liver as they constitute the epithelium lining of the biliary tree, which processes bile production. Some of the first differentiation protocols of hPSCs into hepatocytes reported the presence of bile-duct structures formed by $CK7^+$ and $CK19^+$ cells, which are well known cholangiocyte cell markers [52,53]. However, the first defined protocol for cholangiocyte differentiation from hPSCs was only published several years later [54]. As mentioned before, during hepatic specification when cells become hepatoblasts, they can give rise to both hepatocytes and cholangiocytes. Therefore, existing protocols for cholangiocyte differentiation follow a similar strategy to obtain hepatoblasts. Once this stage is reached, different strategies have been applied for cholangiocyte specification, including the use of epidermal growth factor (EGF) [54,55], FGF10 [56], TGF-β, and activation of NOTCH signaling (either by using Jagged1 or by the incorporation of OP9 stromal cells in a co-culture system) [55,57].

3.3. Other Non-Parenchymal Cells

LSECs are also essential cells of the liver. They are highly specialized endothelial cells that constitute the permeable interface that controls the trafficking of molecules and cells between hepatocytes and the blood stream. Additionally, LSECs play an important role in immunity, liver disease, and regeneration [58,59]. Although important in the maintenance of normal liver functions, there are still few protocols for differentiation of hPSCs into LSECs. In fact, to our best knowledge, there is only one protocol using hPSCs that specifically generates LSECs [60], and a second one using mouse ESCs [61]. Both studies assume that LSECs diverge at some point in ontogeny from endothelial progenitors. Therefore, Koui and colleagues used FLK1$^+$CD31$^+$CD34$^+$ endothelial cells derived from hiPSCs as the starting point to generate and promote a LSEC mature phenotype by adding A83-01 (a TGFβRI inhibitor) under hypoxic conditions. This protocol was able to produce LSECs expressing specific markers like FCGR2B, STAB2, F8, and LYVE1. Generically, many research groups have focused their work on the differentiation of hPSCs into endothelial cells, thus indirectly contributing to future methods for LSEC derivation. For general reviews on this topic, please see [62,63].

Kupffer cells are the liver macrophages, the first line of defense against bacteria, microbial debris, and endotoxins with gastrointestinal origin [64]. In addition to their immunological role, they also cooperate with LSECs in blood clearance [59]. Recently, one study was able to differentiate hiPSCs into Kupffer cells, recapitulating their ontogeny by firstly differentiating hiPSCs into macrophage-precursors and subsequent exposure to hepatic cues by simply culturing these precursors in hepatocyte culture medium (HCM) and advanced DMEM [65].

Finally, HSCs are specialized pericytes that reside in the perisinusoidal space (or space of Disse), between hepatocytes and LSECs. They are important in the maintenance of extracellular matrix homeostasis and its most distinctive feature is the accumulation of vitamin A. Additionally, they play a major role in liver fibrosis upon HSCs activation, i.e., HSCs transdifferentiate from quiescent, vitamin A storing cells into proliferative, fibrogenic myofibroblasts [66,67]. The embryonic origin of HSCs has been and still is elusive, creating a debate around whether these cells originate from the mesoderm, endoderm, or even the neural crest [68]. However, the hypothesis that HSCs originate from the mesoderm (more precisely from the STM) has gained acceptance [69]. This has prompted two studies [60,70] in which hPSCs are firstly differentiated into mesodermal cells, particularly ALCAM$^+$ cell populations, and then divergent paths are used to generate HSCs. Koui and colleagues relied solely on the inhibition of the Rho signaling pathway using Y27632 to achieve an HSC phenotype, whereas Coll and colleagues adopted a more complex strategy adding FGF1, FGF3, palmitic acid, and retinol to the culture. This last study claims that not only cells with phenotypic and functional characteristics of mature HSCs are produced, but also that the method is highly robust with around 78% of PDGFRβ$^+$ cells and 80% vitamin A-storing cells, a feature of mature HSCs.

Table 1. Methods for differentiation of human pluripotent stem cells into hepatic lineages.

Study	Media	Molecules	Ref.
Hepatocytes			
Rambhatla et al., 2003	KO-DMEM+FBS	NaB/DMSO – NaB/HGF	[42]
Kubo et al., 2004	StemPro34 – IMDM+SR	Act A – DEX	[44]
Hay et al., 2008	RPMI+B27 – DMEM+SR+DMSO – L15	Act A/Wnt3a – HGF/OSM	[71]
Song et al., 2009	RPMI – HCM – N2B27	Act A – FGF4/BMP2 – HGF/KGF – OSM/DEX	[43]
Si-Tayeb et al., 2010	RPMI+B27 – HCM	Act A – BMP4/FGF2 – HGF – OSM	[72]
Sullivan et al., 2010	RPMI+B27 – DMEM+SR+DMSO – L15	Act A/Wnt3a – Act A – HGF/OSM	[73]
Touboul et al., 2010	CDM	Act A/Ly/BMP4/FGF2 – FGF10 – FGF10/RA/SB – FGF4/HGF/EGF	[74]
Kajiwara et al., 2012	RPMI+B27 – DMEM+SR+DMSO – HCM	Act A/Wnt3a/NaB – Act A/Wnt3a – HGF/OSM	[75]
Siller et al., 2015	RPMI+B27 – DMEM+SR+DMSO – L15	CHIR – Dihexa/DEX	[48]
Ang et al., 2018	CDM	Act A/CHIR/PI – Act A/LDN – A83/BMP4/FGF2/ATRA – Act A/CHIR/BMP4/Forskolin – BMP4/OSM/DEX/Forsk/Ro/AA/Insulin – DEX/Forskolin/Ro/AA/Insulin	[40]

Table 1. *Cont.*

Study	Media	Molecules	Ref.
Cholangiocytes			
Dianat et al., 2014	RPMI	Act A/Ly – Act A/FGF2/BMP4 – FGF4/HGF/EGF/RA – EGF/GH/IL6	[54]
De Assuncao et al., 2015	RPMI – H69	Act A/Wnt3a – FGF2/BMP4/SHH – SHH/JAG1 – TGFβ	[57]
Ogawa et al., 2015	RPMI – H16 – H16/Ham's F12 – H21/Ham's F12	Act A/CHIR – FGF2/BMP4 – HGF/OSM/DEX – HGF/EGF/TGFβ/OP9	[55]
Sampaziotis et al., 2015	CDM – RPMI – William's E	Act A/Ly/FGF2/BMP4 – Act A – BMP4/SB – Act A/FGF10/RA	[56]
Liver Sinusoidal Endothelial Cells			
Koui et al., 2017	StemPro34 SFM – EGM2	BMP4 – BMP4/Act A/FGF2 – VEGF/SB/Dorsomorphin – VEGF – A83	[60]
Kupffer Cells			
Tasnim et al., 2019	mTeSR1 – X-VIVO – PHCM/Adv DMEM	BMP4/VEGF/SCF – MCSF/IL3 – MCSF	[65]
Hepatic Stellate Cells			
Koui et al., 2017	StemPro34 SFM – MSCGM	BMP4 – BMP4/Act A/FGF2 – VEGF/SB/Dorso – ROCKi	[60]
Coll et al., 2018	MCDB 201	BMP4 – BMP4/FGF1/FGF3 – FGF1/FGF3/PA/Retinol	[70]

The symbol "/" is used to separate media and molecules within the same differentiation step and the symbol "–" is used to separate the different differentiation steps. NaB, sodium butyrate; Act A, Activin A; Ly, Ly294002; SB, SB431542; CHIR, CHIR99021; PI, PI103; LDN, LDN193189; A83, A8301; Ro, Ro4929097; ROCKi, Y-27632.

4. Production of Liver Organoids from Human Pluripotent Stem Cells

The development of advanced 3D culture systems, triggered by the growing knowledge on hPSC differentiation, unlocks the possibility of bringing together multiple organ-specific cell types into a single structure, the so-called organoid [76,77]. Currently, the concept of organoid describes a stem cell-derived 3D structure that through a self-organization process can recapitulate biological features like spatial arrangement, cell–cell, and cell–ECM interactions [9,78]. Accordingly, these structures can provide a much more reliable model of the in vivo anatomy and physiology of a given organ, not only when compared to a 3D cell aggregate composed of a single cell type, but even more when compared to a 2D monolayer culture system. Organoids are thus an emergent system that may serve as building block for tissue engineering applications. The major step forward in this field was given by the groups of Yoshiki Sasai and Hans Clevers in their studies focusing on optic cup [79] and intestinal organoids [80], respectively. Today, scientists can generate organoids from a variety of different cell sources, but some design principles must be taken into consideration when generating an organoid: (1) biophysical properties—the use of solid ECMs to entrap the organoids, or the use of a scaffold-free approach; (2) self-governing of organoid formation—exclusive dependence on endogenous signals or dependence on the addition of exogenous cues; (3) starting cell population—derivation from a single cell, from a homogeneous cell population, or from a co-culture of different cell types [10]. Besides these, recent engineering approaches have been developed to increase the complexity of human organoids and PSCs play an important role in this matter [10,81].

This accumulated knowledge enabled a few groups to recently report the generation of liver organoids derived from hPSCs (Figure 1). The first report of liver organoids dates from 2001 by Michalopoulos and colleagues using several types of adult rat hepatic cells [82], but a more robust and long-term culture of human liver organoids was described later in 2013, using a progenitor population of adult mouse liver [83]. In the same year, Takebe and colleagues were pioneers in using hepatic cells derived from hPSCs, reporting the generation of human liver-like organoids that resemble the developing liver bud during early embryogenesis [84]. These liver bud organoids were generated by mixing hepatic endodermal cells derived from hiPSCs, human umbilical vein endothelial cells (HUVECS) and human mesenchymal stem cells (MSCs). This approach recapitulated the early steps of liver organogenesis resulting in a vascularized human liver bud organoid with improved functionality by producing key liver enzymes. For the maturation of these liver bud organoids, the culture medium used was constituted by HCM and endothelial growth medium (EGM) in 1:1 proportion with the addition of HGF, OSM, and DEX. More recently, the same authors were able to generate fully hPSC-derived liver buds. They used hiPSCs as the cell source to generate hepatic endoderm cells, endothelial cells, and STM cells, and mixed them in a 10:7:2 ratio [85] (Figure 1A). To support the relevance of this approach, complementary in vitro studies demonstrated that besides homotypic interactions between human hepatocytes, heterotypic interactions between hepatocytes and other non-parenchymal cells are critical for self-organization, and that paracrine signals secreted by these cells are important for hepatic maturation [86,87], an idea previously explored in developmental studies using mouse embryos [24]. A similar system was described in 2019 by Pettinato and colleagues using co-cultures of hPSCs with human adipose microvascular endothelial cells (HAMECs) in a 3:1 ratio, that were then submitted to hepatocyte differentiation [88]. This protocol resulted in liver organoids with 89% Albumin$^+$ and 15% CD31$^+$ cells and improved human hepatic functions associated to mature liver cells. Interestingly, HAMECs self-organized in rosette-like structures within the organoids (Figure 1B). Another strategy used 3D aggregates of hepatoblast-like cells derived from hPSCs that were then co-cultured with human fetal liver mesenchymal cells (hFLMCs) [89]. By day 14 of differentiation, Albumin$^+$ cells were found in the peripheral region, whereas PDGFR-β^+ hFLMCs were found in the center of the organoids. Overall, co-culture of hepatoblast-like cells with hFLMCs, in a 2:1 ratio, generated organoids with increased levels of hepatic functions (Figure 1C).

Apart from co-culture of different cell types as a starting point for the generation of liver organoids, other studies have shown how to start with homogeneous cell populations to obtain complex organoids. Since 2017, two different protocols were published using this approach to produce hPSC-derived liver organoids constituted by hepatocytes and cholangiocytes [90,91]. Both studies started with differentiation of hPSCs into hepatoblasts, but they diverge not only in the approach to get to that point, but also in the way they generate liver organoids. Guan and colleagues started with the production of hepatoblast aggregates that with the addition of exogenous growth factors and subsequent dissociation/reaggregation in Matrigel generated liver organoids with both hepatocytes and cholangiocytes (Figure 1D). On the other hand, Wu and colleagues developed a protocol capable of generating liver organoids with 60% ALB$^+$ hepatocytes and about 30% CK19$^+$ cholangiocytes. They claimed that the key factors for the success of their study were the inclusion of 25% mTeSR medium during differentiation and the addition of a cholesterol mixture for organoid functional maturation (Figure 1E). Based on a similar approach, a recent work describes the production of hPSC-derived liver organoids containing hepatocytes, HSCs, Kupffer cells, and cholangiocytes [92]. They initially differentiated hPSCs to foregut, collecting foregut spheroids released from the 2D culture and embedding them in Matrigel with further addition of retinoic acid, a molecule that reportedly plays and important role in the specification not only of parenchymal, but also non-parenchymal liver cells (Figure 1F). Apart from these examples using hPSC-derived cells, other human cell types have been used in the generation of human liver organoids in the past few years [83,93]. Likewise, liver organoids have been produced using mouse [94], rat [95], cat [96], and canine cells [97].

It is important to note that organoid technology is becoming an increasing trend in biomedical research [9]. However, some caution needs to be taken into consideration since by definition, an organoid is a reductionist cell construct that captures the cellular, structural, and physiological complexity of a given organ. Therefore, a clear distinction between 3D spheroids made up of a single cell type and without clear self-organization, and organoids needs to be made. In this section we tried to focus only on reports that fit into this organoid concept.

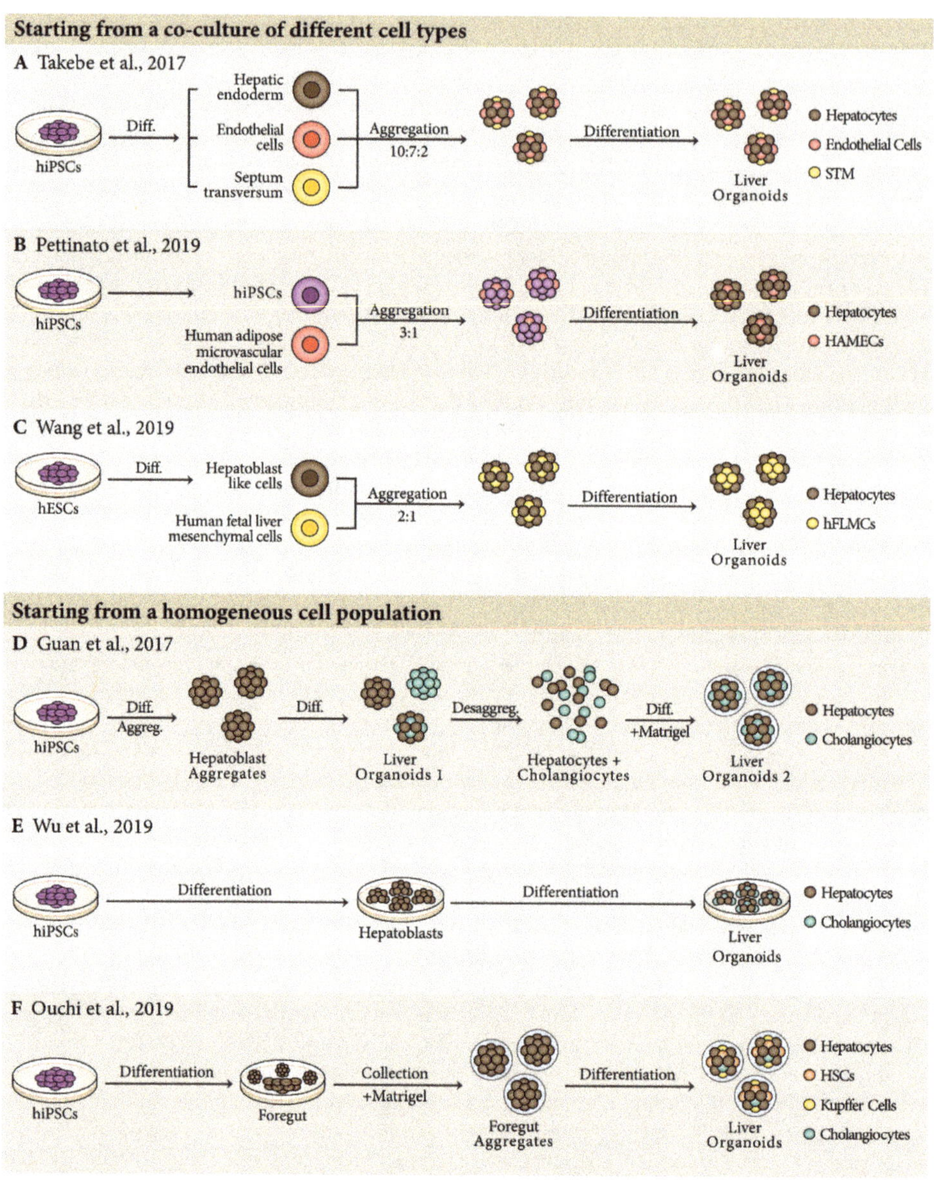

Figure 1. Current strategies for the generation of human pluripotent stem cell (hPSC)-derived liver organoids. So far, liver organoids have been generated by (**A–C**) co-culture of different cell types, including hPSCs, differentiated hepatic cell lineages, or isolated human cells with potential to promote liver organoid differentiation/maturation; (**D–F**) homogeneous cell populations that, through differentiation, are capable of generating cellular constructs with structural and physiological complexity. STM, septum transversum mesenchymal cells; HSCs, hepatic stellate cells.

5. Applications of hPSC-Derived Hepatic Cell Lineages and Liver Organoids

As demonstrated above, hPSCs have the capacity to establish human liver models giving researchers the opportunity to design human liver-based platforms for disease modeling, drug discovery, and hepatotoxicity. Furthermore, differentiated hepatic cells and liver organoids derived from hPSCs represent a renewable source for cell-based therapies aiming at the treatment of patients suffering from liver disease (Figure 2).

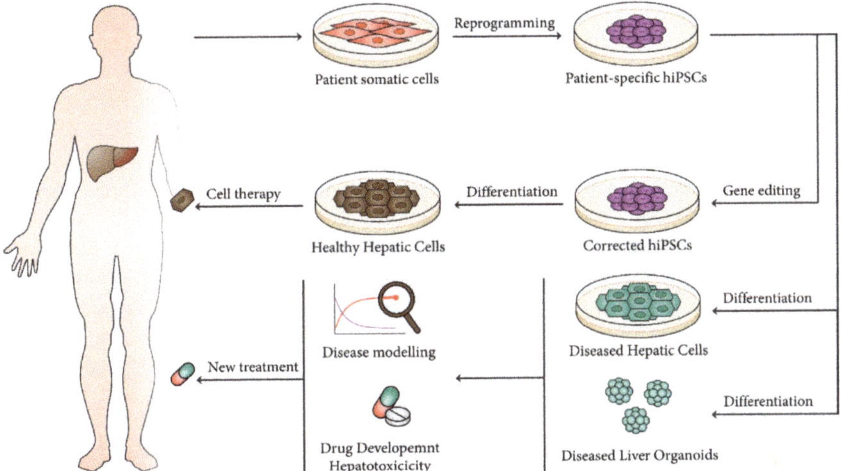

Figure 2. Clinical applications for hPSC-derived hepatic cells and liver organoids. Isolated somatic cells from patients can be cultured and reprogrammed into patient-specific hiPSCs. These cells represent a promising cell source for cell therapy, as differentiated hepatocytes can be used for transplantation in regenerative medicine strategies. Additionally, differentiated hepatic cell lineages or generated liver organoids can be applied in disease modelling, as well as drug development and hepatotoxicity assays.

5.1. Regenerative Medicine

The use of hPSC-derived cells for regeneration or replacement of damaged tissue in regenerative medicine has been proposed to deliver functional recoveries [98]. Indeed, hPSC-based therapies were already established in several clinical trials [99]. hiPSCs in particular have the important advantage of their capability in generating differentiated patient-specific cells, allowing autologous cell transplantation and, theoretically, suppressing the risk of immune rejection. This milestone was achieved in 2014 with the successful transplantation of autologous retinal pigment epithelium sheets derived from hiPSCs [100]. Besides all this progress, chromosomal aberrations (due to cell reprograming and subsequent culture), as well as the tumorigenicity of undifferentiated cells, represent some of the hurdles that need to be taken into consideration when using hPSCs in cell therapies [98,101].

As mentioned above, orthotopic liver transplantation is the single solution for end-stage liver failure and other liver disorders, making this organ the second most common solid transplantation after kidney. Still, given the current transplantation rates, less than 10% of global organ transplantation needs are met [4]. This fact results in the need for alternative therapeutic strategies, and hepatocyte transplantation has been perceived as one [102]. The first attempt to use these cells for transplantation dates back to 1976 using Gunn rats as animal models for Crigler–Najjar syndrome [103]. Still, it was only in 1992 that primary hepatocytes were transplanted into human patients [104]. Since then, numerous patients with liver disease have been treated with hepatocyte transplantation. However, this strategy still presents major hurdles like the limited source of hepatocytes, poor quality of isolated cells, and occasional hepatocyte rejection [102]. Therefore, hPSC-derived hepatocytes have been seen as a potential cell source for transplantation (Table 2). In fact, one of the most successful studies using hPSC-derived hepatocytes for transplantation was able to repopulate up to 15% of the liver of uPA immunodeficient mice after intrasplenic injection [105]. In this study, the differentiated hepatocytes still presented fetal markers like AFP and they were largely negative for isoforms of CYP450. However, after transplantation, the transplanted cell population acquired mature features such as downregulation of AFP expression and cells positive for CYP450 isoforms. This transplantation strategy has also been successful in the alleviation of liver metabolic disorders [106] and in acetaminophen-induced acute toxicity [107], with liver repopulation rates of 2.5–7.5% and 10%, respectively. Thus, hPSC-derived hepatocytes demonstrated their potential to become a relevant cell source for liver cell therapy. However, although initial studies are promising, the use of these cells for regenerative medicine applications needs to be more efficient and effective until it can be translated into human benefit. To accomplished that, recent studies have been using different strategies besides intrasplenic injection for hepatocyte transplantation. One of these strategies is the transplantation of hepatocyte sheets onto the surface of mice livers [108]. Additionally, different scaffolds have been used to support hPSC-derived hepatocytes for subsequent transplantation, namely PCL fibers [109] and decellularized livers [110,111]. Besides this, a recent study reported for the first time, to the best of our knowledge, the use of current good manufacturing practice (cGMP)-compliant hepatocytes generated from hPSCs for transplantation [112]. Liver regeneration will probably require more than simply injecting the right type of cells in the right place. The foundation of knowledge concerning liver regeneration mechanisms, both in normal development and after injury, seems to provide a strong platform to achieve this goal.

Table 2. Summary of recent studies on hPSC-derived hepatocyte transplantation in murine models.

Study	Route	Cells	Nr of Cells	% Repopulation	Ref.
Carpentieret al., 2014	Intrasplenic injection	hPSC-hepatocytes	4×10^6	<1–20%	[105]
Chen et al., 2015	Intrasplenic injection	hPSC-hepatocytes	2×10^6	2.5–7.5%	[106]
Tolosa et al., 2015	Intrasplenic injection	hPSC-hepatocytes	1×10^6	10%	[107]
Nagamoto et al., 2016	Sheet transplantation	hPSC-hepatocyte sheet	8×10^5	-	[108]
Takayama et al., 2017	Intraperitoneal transplantation	hPSC-hepatocytes	1×10^6	-	[113]
Nie et al., 2018	Renal subcapsular space	hPSC-hepatocyte aggregates	1×10^6	-	[114]
Rashidi et al., 2018	Intraperitoneal transplantation	hPSC-hepatocyte aggregates	2×10^6 (aggregates)	-	[109]
	Subcutaneous transplantation	hPSC-hepatocytes in PCL fibers	-	-	[109]
Blackford et al., 2019	Intraperitoneal transplantation	hPSC-hepatocyte aggregates	2×10^3 (aggregates)	-	[112]

5.2. Disease Modeling

In addition to regenerative medicine, disease modeling constitutes another important biomedical application for hPSC derivatives. In vitro disease models based on hPSC technology should improve our knowledge regarding pathological mechanisms underlying human diseases, either genetic or acquired [115]. Given the relevance of liver disease, several studies using hPSC-derived hepatocytes have been published in the last years (for extended review see [7]). However, the effort for improved maturity and greater complexity of in vitro culture systems for disease modeling has led to recent publications using liver organoids as a platform to study liver disease (Table 3).

With the advent of novel gene editing tools like CRISPR-Cas9, it is now possible to induce disease-causing mutations or silencing mutations carried by patient-specific hPSCs and evaluate their effects in differentiated cell phenotypes [116,117]. To understand liver disease mechanisms at the organ level, at least two different studies have been published studying genetic diseases in hPSC-derived liver organoids. Guan and colleagues used patient-specific hPSCs to model Alagille Syndrome (ALGS) and Tetralogy of Fallotthis (TOF) genetic disorders [90]. Firstly, they generated liver organoids from hPSCs reprogrammed from ALGS patients, where in contrast to healthy organoids, mature hepatocytes were developed, but cholangiocytes and bile ductular structure development was impaired. Additionally, they used CRISPR-Cas9 technology to introduce and revert an ALGS causing *JAG1* mutation, C829X, into control and ALGS hPSCs. Thus, ALGS liver pathology was recapitulated, and it was also shown that *JAG1* haploinsufficiency alone does not produce pathology in liver organoids. Moreover, this team also modelled a disease caused by another mutation in *JAG1*, TOF, demonstrating that the type of *JAG1* mutation has a considerable effect in the onset of liver disease. More recently, another study has demonstrated that liver organoids are a suitable platform to model steatohepatitis, a condition that is, among others, characteristic of Wolman disease, caused by a defective activity of lysosomal acid lipase (LAL) [92]. Firstly, these researchers induced steatohepatitis phenotype in liver organoids exposing them to free fatty acids, resulting in lipid accumulation, inflammation, and fibrosis. After that, to highlight the clinical relevance of modelling steatohepatitis, they used patient-derived hPSCs with LAL deficiency to generate liver organoids, thus recapitulating the Wolman disease phenotype with severe steatohepatitis. Additionally, it was demonstrated through liver organoid technology that the steatohepatitis phenotype could be rescued using FGF19, suppressing lipid accumulation and improving liver organoids survival. Besides these two examples of genetic disease modeling, organoids derived from adult liver tissue were already used to study A1AT deficiency and Alagille syndrome [93].

Table 3. Reported studies for disease modeling in hPSC-derived liver organoids.

Study	Disease	Gene/Toxin	Approach	Ref.
Genetic Liver Diseases				
Guan et al., 2017	Alagille syndrome	JAG1	Patient-derived/gene editing	[90]
	Tetralogy of Fallot	JAG1	Patient-derived	[90]
Ouchi et al., 2019	Wolman disease	Free fatty acids	Patient-derived/induced	[92]
Acquired Liver Diseases				
Nie et al., 2018	Hepatitis B	Hepatitis B virus	Induced	[118]
Wang et al., 2019	Alcoholic liver disease	EtOH	Induced	[89]

Recently, liver disease modelling has also been successfully performed to study acquired liver diseases. An example is hepatitis B virus (HBV) infection of hPSC-derived liver organoids [118]. This culture system proved to be more susceptible to HBV when compared to hepatocytes differentiated in a 2D culture system. Particularly, the infection of liver organoids with HBV resulted in hepatic dysfunction with downregulation of hepatic gene expression and emergence of hepatic injury markers, along with the alteration of hepatic structures. Therefore, this study suggested that liver organoids can be considered a good platform for HBV modelling, recapitulating the virus life cycle and consequent dysfunctions. Another example of disease modeling of acquired liver diseases using liver organoids is the study of alcoholic liver disease (ALD), the number one cause of liver-associated mortality in Western countries [89]. Upon EtOH treatment for 7 days, liver organoids displayed liver damage and reduction in cell viability, as well as upregulation of gene expression of fibrogenic markers, thus recapitulating ALD pathophysiology. Additionally, EtOH treatment led to enhanced oxidative stress, an established characteristic of ALD that starts with the metabolism of EtOH by ADH and CYP2E1. Once more, liver organoids proved to be a reliable platform for disease modeling, encouraging its use to study new conditions and eventually contributing to the discovery of new therapeutics.

It is important to note that the cell composition of liver organoids can be of extreme importance when modeling liver diseases. In the examples above, it is possible to understand that given the biliary deficiencies in ALGS and TOF, the presence of cholangiocytes within these organoids it is an essential requirement [90]; similarly, given the characteristic fibrosis of steatohepatitis, HSCs should also be present [92]. Obviously, increasing the complexity of the model system will result in better recreating liver function, and it may even expose the role of the different hepatic cellular components in disease development. In fact, a very recent study shows how the crosstalk between hepatocytes, hepatic Kupffer cells, and HSCs play an important role in alcoholic liver disease (ALD), providing new insights into this pathology and identifying potential new targets for drug therapy [119,120].

5.3. Drug Discovery and Hepatotoxicity

Modeling of human diseases is driven by the need for novel therapeutics aiming at disease treatments and cures. For this reason, drug discovery and toxicological assays are considered a potential application for hPSC derivatives [115,121]. To this end, animal models have been continuously used for drug screening. However, differences between the actual human setting and other animals result in inaccurate prediction of drug effects. Moreover, animal models are not suitable for high-throughput screening of small-molecule libraries [116,122]. As an alternative, the use of hPSC-based models for drug screens have been amply established, assessing not only the efficacy of potential drug candidates, but also their toxicity, predicting the likelihood of potential drugs to cause severe side effects [98]. It is also important to bear in mind that each patient has a specific genetic background, and that this fact implies different responses to medication. Accordingly, hepatocytes and liver organoids generated from hPSCs can be used as a new tool to investigate not only disease mechanisms, but also therapeutic strategies, creating the foundation for personalized therapies, an emerging approach known as precision medicine [122]. Currently, pharmaceutical development is highly costly ($2.6 billion per drug that enters the market) and inefficient (89% of drugs that enter clinical trials will fail due to unforeseen toxicity) [7,123]. Part of this is due to inadequate screening during preclinical studies and so, the use of hPSC-derived hepatocytes and liver organoids in this field is of extreme relevance, as hepatotoxicity is the major type of toxicity associated to drug withdrawals (21% of the cases) [124].

Given this context, there is the urgent need to create protocols that can generate hepatocytes or liver organoids in a scalable and miniaturized fashion, being suitable for high throughput screening of small molecule libraries. Examples of this have been already published [85,125]. These high-throughput screening platforms were already used for identification of drugs for disease treatment with hPSC-derived hepatocytes. At least three studies used small molecules/drug libraries aiming the attenuation or reversion of the effects of diseases like alpha-1-antitrypsin (AAT) deficiency [126], familial hypercholesterolemia [127], and mitochondrial DNA depletion syndrome (MTDPS3) [128]. The same platforms have also been used for toxicity screens evaluating the effect of certain drugs on hPSC-derived hepatocytes, typically testing compounds known to be toxic and non-toxic and assessing cell morphology and viability [129–131].

All of these studies are mainly focused on hepatocytes, but as described before non-parenchymal cells hold great importance in liver physiology. For instance, LSECs are implicated in most liver diseases making them an attractive therapeutic target [58]. Additionally, Kupffer cells play a crucial role in drug-induced liver injury (DILI) and other liver diseases. To demonstrate this, hepatocytes have been co-cultured with Kupffer cells resulting in a model that is more sensitive in detecting hepatotoxicity induced by different drugs [65]. Thus, the development of more complex liver organoids composed of different hepatic cell types can substantially benefit drug discovery and hepatotoxicity assays.

Drug screening can also be performed using microfluidic devices like the so-called organ-on-a-chip or microphysiological systems (MPSs), where living cells can be cultured in continuously perfused chambers, modeling the physiological functions of a given tissue or organ [132,133]. Accordingly, organ-on-a-chip is also a valuable platform for drug development and toxicology giving insights into adsorption, distribution, metabolism, elimination, and toxicity (ADMET), mathematical pharmacokinetics (PK), pharmacodynamics (PD), and drug efficacy [134]. In fact, Wang and colleagues have recently achieved the in situ differentiation of hPSCs into liver organoids using a perfusable micropillar chip [135]. The on-chip liver organoids displayed both hepatocytes and cholangiocytes, as well as increased cell viability and mature cell signature. Notably, the organoids generated in this platform not only presented high levels of cytochrome P450 enzyme expression, but also dose- and time-dependent hepatotoxic response to acetaminophen. These results support the notion that organ-on-a-chip technology constitutes a valid platform for drug testing. Moreover, organ-on-a-chip technology can rely not only on individual designs, but also in more complex interlinked multi-organ-on-chips, or body-on-a-chip platforms capable of mimicking multi-organ crosstalk [136]. Indeed, the study of an inter-tissue crosstalk between gut and liver during inflammatory processes was already reported [137], and more recently an interconnected MPS representing up to 10 organs, including liver, was established [138]. Applying this technology to hPSC-derivatives is still, to the best of our knowledge, an unmet need.

6. Conclusions

Stem cell research has made considerable progress in the last years, providing researchers with the necessary tools to replicate in vitro the developmental processes from hepatogenesis and, ultimately, the physiological and structural features of human liver. To that end, numerous strategies have been published for hPSC direct differentiation into hepatocytes, cholangiocytes, and other non-parenchymal hepatic cell types. However, some of these cells still present a fetal phenotype with only a certain degree of maturity. Recently, organoid technology has contributed to the increasing complexity of hepatic culture systems in vitro, leading to enhanced metabolic functionality and cellular architecture, and furthering our knowledge of liver biology. As new technologies are being developed, the potential applications of hepatic cell lineages and liver organoids derived from hPSCs are increasing, bringing new insights into the fields of regenerative medicine, disease modelling, drug discovery, and hepatotoxicity. Thus, it is in the intersection of biology and engineering that many open questions concerning the human liver can be brought to light, leading, eventually, to an improvement in the human condition.

Author Contributions: J.P.C. and T.G.F. contributed to the design, to the writing, and to the review of the manuscript. All authors have read and agreed to the published version of the manuscript.

Funding: This work was supported by Fundação para a Ciência e a Tecnologia (FCT), Portugal (UIDB/04565/2020) through Programa Operacional Regional de Lisboa 2020 (Project N. 007317), and projects co-funded by FEDER (POR Lisboa 2020—Programa Operacional Regional de Lisboa PORTUGAL 2020) and FCT through grant PAC-PRECISE LISBOA-01-0145-FEDER-016394 and CEREBEX Generation of Cerebellar Organoids for Ataxia Research grant LISBOA-01-0145-FEDER-029298.

Acknowledgments: J.P.C. acknowledges Fundação para a Ciência e Tecnologia (FCT, Portugal, http://www.fct.pt) for financial support (SFRH/PD/BD135500/2018).

Conflicts of Interest: The authors declare no conflict of interest.

References

1. Si-Tayeb, K.; Lemaigre, F.P.; Duncan, S.A. Organogenesis and Development of the Liver. *Dev. Cell* **2010**, *18*, 175–189. [CrossRef] [PubMed]
2. Miyajima, A.; Tanaka, M.; Itoh, T. Stem/progenitor cells in liver development, homeostasis, regeneration, and reprogramming. *Cell Stem Cell* **2014**, *14*, 561–574. [CrossRef] [PubMed]
3. Bhatia, S.N.; Underhill, G.H.; Zaret, K.S.; Fox, I.J. Cell and tissue engineering for liver disease. *Sci. Transl. Med.* **2014**, *6*, 245sr2. [CrossRef] [PubMed]
4. Asrani, S.K.; Devarbhavi, H.; Eaton, J.; Kamath, P.S. Burden of liver diseases in the world. *J. Hepatol.* **2019**, *70*, 151–171. [CrossRef] [PubMed]
5. Schwartz, R.E.; Fleming, H.E.; Khetani, S.R.; Bhatia, S.N. Pluripotent stem cell-derived hepatocyte-like cells. *Biotechnol. Adv.* **2014**, *32*, 504–513. [CrossRef] [PubMed]
6. Hannoun, Z.; Steichen, C.; Dianat, N.; Weber, A.; Dubart-Kupperschmitt, A. The potential of induced pluripotent stem cell derived hepatocytes. *J. Hepatol.* **2016**, *65*, 182–199. [CrossRef]
7. Corbett, J.L.; Duncan, S.A. iPSC-Derived Hepatocytes as a Platform for Disease Modeling and Drug Discovery. *Front. Med.* **2019**, *6*, 1–12. [CrossRef]
8. Aizarani, N.; Saviano, A.; Sagar; Mailly, L.; Durand, S.; Herman, J.S.; Pessaux, P.; Baumert, T.F.; Grün, D. A human liver cell atlas reveals heterogeneity and epithelial progenitors. *Nature* **2019**, *572*, 199–204. [CrossRef]

9. Rossi, G.; Manfrin, A.; Lutolf, M.P. Progress and potential in organoid research. *Nat. Rev. Genet.* **2018**, *19*, 671–687. [CrossRef]
10. Cotovio, J.P.; Fernandes, T.G.; Diogo, M.M.; Cabral, J.M.S. Pluripotent stem cell biology and engineering. In *Engineering Strategies for Regenerative Medicine*; Elsevier: Amsterdam, The Netherlands, 2020; pp. 1–31.
11. Lawson, K.A.; Meneses, J.J.; Pedersen, R.A. Clonal analysis of epiblast fate during germ layer formation in the mouse embryo. *Development* **1991**, *113*, 891–911.
12. Burdsal, C.A.; Damsky, C.H.; Pedersen, R.A. The role of E-cadherin and integrins in mesoderm differentiation and migration at the mammalian primitive streak. *Development* **1993**, *118*, 829–844. [PubMed]
13. Kwon, G.S.; Viotti, M.; Hadjantonakis, A.K. The Endoderm of the Mouse Embryo Arises by Dynamic Widespread Intercalation of Embryonic and Extraembryonic Lineages. *Dev. Cell* **2008**, *15*, 509–520. [CrossRef] [PubMed]
14. Viotti, M.; Nowotschin, S.; Hadjantonakis, A.K. SOX17 links gut endoderm morphogenesis and germ layer segregation. *Nat. Cell Biol.* **2014**, *16*, 1146–1156. [CrossRef] [PubMed]
15. Lowe, L.A.; Yamada, S.; Kuehn, M.R. Genetic dissection of nodal function in patterning the mouse embryo. *Development* **2001**, *128*, 1831–1843.
16. Vincent, S.D.; Dunn, N.R.; Hayashi, S.; Norris, D.P.; Robertson, E.J. Cell fate decisions within the mouse organizer are governed by graded Nodal signals. *Genes Dev.* **2003**, *17*, 1646–1662. [CrossRef]
17. Faial, T.; Bernardo, A.S.; Mendjan, S.; Diamanti, E.; Ortmann, D.; Gentsch, G.E.; Mascetti, V.L.; Trotter, M.W.B.; Smith, J.C.; Pedersen, R.A. Brachyury and SMAD signalling collaboratively orchestrate distinct mesoderm and endoderm gene regulatory networks in differentiating human embryonic stem cells. *Development* **2015**, *142*, 2121–2135. [CrossRef]
18. Tremblay, K.D.; Zaret, K.S. Distinct populations of endoderm cells converge to generate the embryonic liver bud and ventral foregut tissues. *Dev. Biol.* **2005**, *280*, 87–99. [CrossRef]
19. Gordillo, M.; Evans, T.; Gouon-Evans, V. Orchestrating liver development. *Development* **2015**, *142*, 2094–2108. [CrossRef]
20. Gualdi, R.; Bossard, P.; Zheng, M.; Hamada, Y.; Coleman, J.R.; Zaret, K.S. Hepatic specification of the gut endoderm in vitro: cell signaling and transcriptional control. *Genes Dev.* **1996**, *10*, 1670–1682. [CrossRef]
21. Rossi, J.M.; Dunn, N.R.; Hogan, B.L.M.; Zaret, K.S. Distinct mesodermal signals, including BMPs from the septum, transversum mesenchyme, are required in combination for hepatogenesis from the endoderm. *Genes Dev.* **2001**, *15*, 1998–2009. [CrossRef]
22. Serls, A.E.; Doherty, S.; Parvatiyar, P.; Wells, J.M.; Deutsch, G.H. Different thresholds of fibroblast growth factors pattern the ventral foregut into liver and lung. *Development* **2005**, *132*, 35–47. [CrossRef] [PubMed]
23. Calmont, A.; Wandzioch, E.; Tremblay, K.D.; Minowada, G.; Kaestner, K.H.; Martin, G.R.; Zaret, K.S. An FGF Response Pathway that Mediates Hepatic Gene Induction in Embryonic Endoderm Cells. *Dev. Cell* **2006**, *11*, 339–348. [CrossRef] [PubMed]
24. Matsumoto, K.; Yoshitomi, H.; Rossant, J.; Zaret, K.S. Liver organogenesis promoted by endothelial cells prior to vascular function. *Science* **2001**, *294*, 559–563. [CrossRef] [PubMed]
25. Bort, R.; Signore, M.; Tremblay, K.; Pedro, J.; Barbera, M.; Zaret, K.S. Hex homeobox gene controls the transition of the endoderm to a pseudostratified, cell emergent epithelium for liver bud development. *Dev. Biol.* **2006**, *290*, 44–56. [CrossRef] [PubMed]
26. Sosa-Pineda, B.; Wigle, J.T.; Oliver, G. Hepatocyte migration during liver development requires Prox1. *Nat. Genet.* **2000**, *25*, 254–255. [CrossRef] [PubMed]
27. Lüdtke, T.H.W.; Christoffels, V.M.; Petry, M.; Kispert, A. Tbx3 promotes liver bud expansion during mouse development by suppression of cholangiocyte differentiation. *Hepatology* **2009**, *49*, 969–978. [CrossRef]
28. Parviz, F.; Matullo, C.; Garrison, W.D.; Savatski, L.; Adamson, J.W.; Ning, G.; Kaestner, K.H.; Rossi, J.M.; Zaret, K.S.; Duncan, S.A. Hepatocyte nuclear factor 4α controls the development of a hepatic epithelium and liver morphogenesis. *Nat. Genet.* **2003**, *34*, 292–296. [CrossRef]
29. Clotman, F.; Jacquemin, P.; Plumb-Rudewiez, N.; Pierreux, C.E.; Van der Smissen, P.; Dietz, H.C.; Courtoy, P.J.; Rousseau, G.G.; Lemaigre, F.P. Control of liver cell fate decision by a gradient of TGF beta signaling modulated by Onecut transcription factors. *Genes Dev.* **2005**, *19*, 1849–1854. [CrossRef]
30. Ober, E.A.; Lemaigre, F.P. Development of the liver: Insights into organ and tissue morphogenesis. *J. Hepatol.* **2018**, *68*, 1049–1062. [CrossRef]

31. Evans, M.J.; Kaufman, M.H. Establishment in culture of pluripotential cells from mouse embryos. *Nature* **1981**, *292*, 154–156. [CrossRef]
32. Martin, G.R. Isolation of a pluripotent cell line from early mouse embryos cultured in medium conditioned by teratocarcinoma stem cells. *Proc. Natl. Acad. Sci. USA* **1981**, *78*, 7634–7638. [CrossRef] [PubMed]
33. Thomson, J.A. Embryonic Stem Cell Lines Derived from Human Blastocysts. *Science* **1998**, *282*, 1145–1147. [CrossRef] [PubMed]
34. Takahashi, K.; Yamanaka, S. Induction of pluripotent stem cells from mouse embryonic and adult fibroblast cultures by defined factors. *Cell* **2006**, *126*, 663–676. [CrossRef] [PubMed]
35. Takahashi, K.; Tanabe, K.; Ohnuki, M.; Narita, M.; Ichisaka, T.; Tomoda, K.; Yamanaka, S. Induction of Pluripotent Stem Cells from Adult Human Fibroblasts by Defined Factors. *Cell* **2007**, *131*, 861–872. [CrossRef] [PubMed]
36. Yu, J.; Vodyanik, M.; Smuga-Otto, K.; Antosiewicz-Bourget, J.; Frane, J.; Tian, S.; Nie, J.; Jonsdottir, G.; Ruotti, V.; Stewart, R.; et al. Induced Pluripotent Stem Cell Lines Derived from Human Somatic Cells. *Science* **2007**, *318*, 1917–1920. [CrossRef]
37. Chambers, S.M.; Fasano, C.A.; Papapetrou, E.P.; Tomishima, M.; Sadelain, M.; Studer, L. Highly efficient neural conversion of human ES and iPS cells by dual inhibition of SMAD signaling. *Nat. Biotechnol.* **2009**, *27*, 275–280. [CrossRef]
38. Branco, M.A.; Cotovio, J.P.; Rodrigues, C.A.V.; Vaz, S.H.; Fernandes, T.G.; Moreira, L.M.; Cabral, J.M.S.; Diogo, M.M. Transcriptomic analysis of 3D Cardiac Differentiation of Human Induced Pluripotent Stem Cells Reveals Faster Cardiomyocyte Maturation Compared to 2D Culture. *Sci. Rep.* **2019**, *9*, 1–13. [CrossRef]
39. Bellin, M.; Marchetto, M.C.; Gage, F.H.; Mummery, C.L. Induced pluripotent stem cells: The new patient? *Nat. Rev. Mol. Cell Biol.* **2012**, *13*, 713–726. [CrossRef]
40. Ang, L.T.; Tan, A.K.Y.; Autio, M.I.; Goh, S.H.; Choo, S.H.; Lee, K.L.; Tan, J.; Pan, B.; Lee, J.J.H.; Lum, J.J.; et al. A Roadmap for Human Liver Differentiation from Pluripotent Stem Cells. *Cell Rep.* **2018**, *22*, 2190–2205. [CrossRef]
41. Abe, K.; Niwa, H.; Iwase, K.; Takiguchi, M.; Mori, M.; Abé, S.I.; Abe, K.; Yamamura, K.I. Endoderm-specific gene expression in embryonic stem cells differentiated to embryoid bodies. *Exp. Cell Res.* **1996**, *229*, 27–34. [CrossRef]
42. Rambhatla, L.; Chiu, C.P.; Kundu, P.; Peng, Y.; Carpenter, M.K. Generation of hepatocyte-like cells from human embryonic stem cells. *Cell Transplant.* **2003**, *12*, 1–11. [CrossRef] [PubMed]
43. Song, Z.; Cai, J.; Liu, Y.; Zhao, D.; Yong, J.; Duo, S.; Song, X.; Guo, Y.; Zhao, Y.; Qin, H.; et al. Efficient generation of hepatocyte-like cells from human induced pluripotent stem cells. *Cell Res.* **2009**, *19*, 1233–1242. [CrossRef] [PubMed]
44. Kubo, A.; Shinozaki, K.; Shannon, J.M.; Kouskoff, V.; Kennedy, M.; Woo, S.; Fehling, H.J.; Keller, G. Development of definitive endoderm from embryonic stem cells in culture. *Development* **2004**, *131*, 1651–1662. [CrossRef]
45. Chen, C.; Soto-Gutierrez, A.; Baptista, P.M.; Spee, B. Biotechnology Challenges to In Vitro Maturation of Hepatic Stem Cells. *Gastroenterology* **2018**, *154*, 1258–1272. [CrossRef] [PubMed]
46. Kamiya, A.; Kinoshita, T.; Miyajima, A. Oncostatin M and hepatocyte growth factor induce hepatic maturation via distinct signaling pathways. *FEBS Lett.* **2001**, *492*, 90–94. [CrossRef]
47. Kamiya, A. Fetal liver development requires a paracrine action of oncostatin M through the gp130 signal transducer. *EMBO J.* **1999**, *18*, 2127–2136. [CrossRef]
48. Siller, R.; Greenhough, S.; Naumovska, E.; Sullivan, G.J. Small-molecule-driven hepatocyte differentiation of human pluripotent stem cells. *Stem Cell Rep.* **2015**, *4*, 939–952. [CrossRef]
49. Vosough, M.; Omidinia, E.; Kadivar, M.; Shokrgozar, M.A.; Pournasr, B.; Aghdami, N.; Baharvand, H. Generation of functional hepatocyte-like cells from human pluripotent stem cells in a scalable suspension culture. *Stem Cells Dev.* **2013**, *22*, 2693–2705. [CrossRef]
50. Farzaneh, Z.; Najarasl, M.; Abbasalizadeh, S.; Vosough, M.; Baharvand, H. Developing a Cost-Effective and Scalable Production of Human Hepatic Competent Endoderm from Size-Controlled Pluripotent Stem Cell Aggregates. *Stem Cells Dev.* **2018**, *27*, 262–274. [CrossRef]
51. Yamashita, T.; Takayama, K.; Sakurai, F.; Mizuguchi, H. Billion-scale production of hepatocyte-like cells from human induced pluripotent stem cells. *Biochem. Biophys. Res. Commun.* **2018**, *496*, 1269–1275. [CrossRef]

52. Cai, J.; Zhao, Y.; Liu, Y.; Ye, F.; Song, Z.; Qin, H.; Meng, S.; Chen, Y.; Zhou, R.; Song, X.; et al. Directed differentiation of human embryonic stem cells into functional hepatic cells. *Hepatology* **2007**, *45*, 1229–1239. [CrossRef] [PubMed]
53. Zhao, D.; Chen, S.; Cai, J.; Guo, Y.; Song, Z.; Che, J.; Liu, C.; Wu, C.; Ding, M.; Deng, H. Derivation and characterization of hepatic progenitor cells from human embryonic stem cells. *PLoS ONE* **2009**, *4*. [CrossRef] [PubMed]
54. Dianat, N.; Dubois-Pot-Schneider, H.; Steichen, C.; Desterke, C.; Leclerc, P.; Raveux, A.; Combettes, L.; Weber, A.; Corlu, A.; Dubart-Kupperschmitt, A. Generation of functional cholangiocyte-like cells from human pluripotent stem cells and HepaRG cells. *Hepatology* **2014**, *60*, 700–714. [CrossRef]
55. Ogawa, M.; Ogawa, S.; Bear, C.E.; Ahmadi, S.; Chin, S.; Li, B.; Grompe, M.; Keller, G.; Kamath, B.M.; Ghanekar, A. Directed differentiation of cholangiocytes from human pluripotent stem cells. *Nat. Biotechnol.* **2015**, *33*, 853–861. [CrossRef] [PubMed]
56. Sampaziotis, F.; De Brito, M.C.; Madrigal, P.; Bertero, A.; Saeb-Parsy, K.; Soares, F.A.C.; Schrumpf, E.; Melum, E.; Karlsen, T.H.; Bradley, J.A.; et al. Cholangiocytes derived from human induced pluripotent stem cells for disease modeling and drug validation. *Nat. Biotechnol.* **2015**, *33*, 845–852. [CrossRef] [PubMed]
57. De Assuncao, T.M.; Sun, Y.; Jalan-Sakrikar, N.; Drinane, M.C.; Huang, B.Q.; Li, Y.; Davila, J.I.; Wang, R.; O'Hara, S.P.; Lomberk, G.A.; et al. Development and characterization of human-induced pluripotent stem cell-derived cholangiocytes. *Lab. Investig.* **2015**, *95*, 684–696. [CrossRef]
58. Poisson, J.; Lemoinne, S.; Boulanger, C.; Durand, F.; Moreau, R.; Valla, D.; Rautou, P.E. Liver sinusoidal endothelial cells: Physiology and role in liver diseases. *J. Hepatol.* **2017**, *66*, 212–227. [CrossRef]
59. Sørensen, K.K.; Simon-Santamaria, J.; McCuskey, R.S.; Smedsrød, B. Liver sinusoidal endothelial cells. *Compr. Physiol.* **2015**, *5*, 1751–1774.
60. Koui, Y.; Kido, T.; Ito, T.; Oyama, H.; Chen, S.W.; Katou, Y.; Shirahige, K.; Miyajima, A. An In Vitro Human Liver Model by iPSC-Derived Parenchymal and Non-parenchymal Cells. *Stem Cell Rep.* **2017**, *9*, 490–498. [CrossRef]
61. Arai, T.; Sakurai, T.; Kamiyoshi, A.; Ichikawa-Shindo, Y.; Iinuma, N.; Iesato, Y.; Koyama, T.; Yoshizawa, T.; Uetake, R.; Yamauchi, A.; et al. Induction of LYVE-1/stabilin-2-positive liver sinusoidal endothelial-like cells from embryoid bodies by modulation of adrenomedullin-RAMP2 signaling. *Peptides* **2011**, *32*, 1855–1865. [CrossRef]
62. Williams, I.M.; Wu, J.C. Generation of Endothelial Cells From Human Pluripotent Stem Cells. *Arterioscler. Thromb. Vasc. Biol.* **2019**, *39*, 1317–1329. [CrossRef] [PubMed]
63. Xu, M.; He, J.; Zhang, C.; Xu, J.; Wang, Y. Strategies for derivation of endothelial lineages from human stem cells. *Stem Cell Res. Ther.* **2019**, *10*, 1–14. [CrossRef] [PubMed]
64. Li, P.; He, K.; Li, J.; Liu, Z.; Gong, J. The role of Kupffer cells in hepatic diseases. *Mol. Immunol.* **2017**, *85*, 222–229. [CrossRef] [PubMed]
65. Tasnim, F.; Xing, J.; Huang, X.; Mo, S.; Wei, X.; Tan, M.H.; Yu, H. Generation of mature kupffer cells from human induced pluripotent stem cells. *Biomaterials* **2019**, *192*, 377–391. [CrossRef] [PubMed]
66. Tsuchida, T.; Friedman, S.L. Mechanisms of hepatic stellate cell activation. *Nat. Rev. Gastroenterol. Hepatol.* **2017**, *14*, 397–411. [CrossRef] [PubMed]
67. Higashi, T.; Friedman, S.L.; Hoshida, Y. Hepatic stellate cells as key target in liver fibrosis. *Adv. Drug Deliv. Rev.* **2017**, *121*, 27–42. [CrossRef] [PubMed]
68. Friedman, S.L. Hepatic stellate cells: Protean, multifunctional, and enigmatic cells of the liver. *Physiol. Rev.* **2008**, *88*, 125–172. [CrossRef]
69. Asahina, K.; Zhou, B.; Pu, W.T.; Tsukamoto, H. Septum transversum-derived mesothelium gives rise to hepatic stellate cells and perivascular mesenchymal cells in developing mouse liver. *Hepatology* **2011**, *53*, 983–995. [CrossRef]
70. Coll, M.; Perea, L.; Boon, R.; Leite, S.B.; Vallverdú, J.; Mannaerts, I.; Smout, A.; El Taghdouini, A.; Blaya, D.; Rodrigo-Torres, D.; et al. Generation of Hepatic Stellate Cells from Human Pluripotent Stem Cells Enables In Vitro Modeling of Liver Fibrosis. *Cell Stem Cell* **2018**, *23*, 101–113.e7. [CrossRef]
71. Hay, D.C.; Fletcher, J.; Payne, C.; Terrace, J.D.; Gallagher, R.C.J.; Snoeys, J.; Black, J.R.; Wojtacha, D.; Samuel, K.; Hannoun, Z.; et al. Highly efficient differentiation of hESCs to functional hepatic endoderm requires ActivinA and Wnt3a signaling. *Proc. Natl. Acad. Sci. USA* **2008**, *105*, 12301–12306. [CrossRef]

72. Si-Tayeb, K.; Noto, F.K.; Nagaoka, M.; Li, J.; Battle, M.A.; Duris, C.; North, P.E.; Dalton, S.; Duncan, S.A. Highly efficient generation of human hepatocyte-like cells from induced pluripotent stem cells. *Hepatology* **2010**, *51*, 297–305. [CrossRef] [PubMed]
73. Sullivan, G.J.; Hay, D.C.; Park, I.-H.; Fletcher, J.; Hannoun, Z.; Payne, C.M.; Dalgetty, D.; Black, J.R.; Ross, J.A.; Samuel, K.; et al. Generation of functional human hepatic endoderm from human induced pluripotent stem cells. *Hepatology* **2010**, *51*, 329–335. [CrossRef] [PubMed]
74. Touboul, T.; Hannan, N.R.F.; Corbineau, S.; Martinez, A.; Martinet, C.; Branchereau, S.; Mainot, S.; Strick-Marchand, H.; Pedersen, R.; Di Santo, J.; et al. Generation of functional hepatocytes from human embryonic stem cells under chemically defined conditions that recapitulate liver development. *Hepatology* **2010**, *51*, 1754–1765. [CrossRef] [PubMed]
75. Kajiwara, M.; Aoi, T.; Okita, K.; Takahashi, R.; Inoue, H.; Takayama, N.; Endo, H.; Eto, K.; Toguchida, J.; Uemoto, S.; et al. Donor-dependent variations in hepatic differentiation from human-induced pluripotent stem cells. *Proc. Natl. Acad. Sci. USA* **2012**, *109*, 12538–12543. [CrossRef] [PubMed]
76. Fatehullah, A.; Tan, S.H.; Barker, N. Organoids as an in vitro model of human development and disease. *Nat. Cell Biol.* **2016**, *18*, 246–254. [CrossRef] [PubMed]
77. Lancaster, M.A.; Knoblich, J.A. Organogenesis in a dish: modeling development and disease using organoid technologies. *Science* **2014**, *345*, 1247125. [CrossRef]
78. Yin, X.; Mead, B.E.; Safaee, H.; Langer, R.; Karp, J.M.; Levy, O. Engineering Stem Cell Organoids. *Cell Stem Cell* **2016**, *18*, 25–38. [CrossRef]
79. Eiraku, M.; Takata, N.; Ishibashi, H.; Kawada, M.; Sakakura, E.; Okuda, S.; Sekiguchi, K.; Adachi, T.; Sasai, Y. Self-organizing optic-cup morphogenesis in three-dimensional culture. *Nature* **2011**, *472*, 51–56. [CrossRef]
80. Sato, T.; Vries, R.G.; Snippert, H.J.; van de Wetering, M.; Barker, N.; Stange, D.E.; van Es, J.H.; Abo, A.; Kujala, P.; Peters, P.J.; et al. Single Lgr5 stem cells build crypt-villus structures in vitro without a mesenchymal niche. *Nature* **2009**, *459*, 262–265. [CrossRef]
81. Silva, T.P.; Cotovio, J.P.; Bekman, E.; Carmo-Fonseca, M.; Cabral, J.M.S.; Fernandes, T.G. Design Principles for Pluripotent Stem Cell-Derived Organoid Engineering. *Stem Cells Int.* **2019**, *2019*, 1–17. [CrossRef]
82. Michalopoulos, G.K.; Bowen, W.C.; Mulè, K.; Stolz, D.B. Histological organization in hepatocyte organoid cultures. *Am. J. Pathol.* **2001**, *159*, 1877–1887. [CrossRef]
83. Huch, M.; Dorrell, C.; Boj, S.F.; Van Es, J.H.; Li, V.S.W.; Van De Wetering, M.; Sato, T.; Hamer, K.; Sasaki, N.; Finegold, M.J.; et al. In vitro expansion of single Lgr5 + liver stem cells induced by Wnt-driven regeneration. *Nature* **2013**, *494*, 247–250. [CrossRef] [PubMed]
84. Takebe, T.; Sekine, K.; Enomura, M.; Koike, H.; Kimura, M.; Ogaeri, T.; Zhang, R.R.; Ueno, Y.; Zheng, Y.W.; Koike, N.; et al. Vascularized and functional human liver from an iPSC-derived organ bud transplant. *Nature* **2013**, *499*, 481–484. [CrossRef] [PubMed]
85. Takebe, T.; Sekine, K.; Kimura, M.; Yoshizawa, E.; Ayano, S.; Koido, M.; Funayama, S.; Nakanishi, N.; Hisai, T.; Kobayashi, T.; et al. Massive and Reproducible Production of Liver Buds Entirely from Human Pluripotent Stem Cells. *Cell Rep.* **2017**, *21*, 2661–2670. [CrossRef] [PubMed]
86. Camp, J.G.; Sekine, K.; Gerber, T.; Loeffler-Wirth, H.; Binder, H.; Gac, M.; Kanton, S.; Kageyama, J.; Damm, G.; Seehofer, D.; et al. Multilineage communication regulates human liver bud development from pluripotency. *Nature* **2017**, *546*, 533–538. [CrossRef] [PubMed]
87. Asai, A.; Aihara, E.; Watson, C.; Mourya, R.; Mizuochi, T.; Shivakumar, P.; Phelan, K.; Mayhew, C.; Helmrath, M.; Takebe, T.; et al. Paracrine signals regulate human liver organoid maturation from induced pluripotent stem cells. *Development* **2017**, *144*, 1056–1064. [CrossRef]
88. Pettinato, G.; Lehoux, S.; Ramanathan, R.; Salem, M.M.; He, L.X.; Muse, O.; Flaumenhaft, R.; Thompson, M.T.; Rouse, E.A.; Cummings, R.D.; et al. Generation of fully functional hepatocyte-like organoids from human induced pluripotent stem cells mixed with Endothelial Cells. *Sci. Rep.* **2019**, *9*, 1–21. [CrossRef]
89. Wang, S.; Wang, X.; Tan, Z.; Su, Y.; Liu, J.; Chang, M.; Yan, F.; Chen, J.; Chen, T.; Li, C.; et al. Human ESC-derived expandable hepatic organoids enable therapeutic liver repopulation and pathophysiological modeling of alcoholic liver injury. *Cell Res.* **2019**, *29*, 1009–1026. [CrossRef]
90. Guan, Y.; Xu, D.; Garfin, P.M.; Ehmer, U.; Hurwitz, M.; Enns, G.; Michie, S.; Wu, M.; Zheng, M.; Nishimura, T.; et al. Human hepatic organoids for the analysis of human genetic diseases. *JCI Insight* **2017**, *2*. [CrossRef]

91. Wu, F.; Wu, D.; Ren, Y.; Huang, Y.; Feng, B.; Zhao, N.; Zhang, T.; Chen, X.; Chen, S.; Xu, A. Generation of hepatobiliary organoids from human induced pluripotent stem cells. *J. Hepatol.* **2019**, *70*, 1145–1158. [CrossRef]
92. Ouchi, R.; Togo, S.; Kimura, M.; Shinozawa, T.; Koido, M.; Koike, H.; Thompson, W.; Karns, R.A.; Mayhew, C.N.; McGrath, P.S.; et al. Modeling Steatohepatitis in Humans with Pluripotent Stem Cell-Derived Organoids. *Cell Metab.* **2019**, *30*, 374–384.e6. [CrossRef] [PubMed]
93. Huch, M.; Gehart, H.; Van Boxtel, R.; Hamer, K.; Blokzijl, F.; Verstegen, M.M.A.; Ellis, E.; Van Wenum, M.; Fuchs, S.A.; De Ligt, J.; et al. Long-term culture of genome-stable bipotent stem cells from adult human liver. *Cell* **2015**, *160*, 299–312. [CrossRef] [PubMed]
94. Hu, H.; Gehart, H.; Artegiani, B.; LÖpez-Iglesias, C.; Dekkers, F.; Basak, O.; van Es, J.; Chuva de Sousa Lopes, S.M.; Begthel, H.; Korving, J.; et al. Long-Term Expansion of Functional Mouse and Human Hepatocytes as 3D Organoids. *Cell* **2018**, *175*, 1591–1606.e19. [CrossRef] [PubMed]
95. Kuijk, E.W.; Rasmussen, S.; Blokzijl, F.; Huch, M.; Gehart, H.; Toonen, P.; Begthel, H.; Clevers, H.; Geurts, A.M.; Cuppen, E. Generation and characterization of rat liver stem cell lines and their engraftment in a rat model of liver failure. *Sci. Rep.* **2016**, *6*, 1–11. [CrossRef] [PubMed]
96. Kruitwagen, H.S.; Oosterhoff, L.A.; Vernooij, I.G.W.H.; Schrall, I.M.; van Wolferen, M.E.; Bannink, F.; Roesch, C.; van Uden, L.; Molenaar, M.R.; Helms, J.B.; et al. Long-Term Adult Feline Liver Organoid Cultures for Disease Modeling of Hepatic Steatosis. *Stem Cell Rep.* **2017**, *8*, 822–830. [CrossRef]
97. Nantasanti, S.; Spee, B.; Kruitwagen, H.S.; Chen, C.; Geijsen, N.; Oosterhoff, L.A.; Van Wolferen, M.E.; Pelaez, N.; Fieten, H.; Wubbolts, R.W.; et al. Disease modeling and gene therapy of copper storage disease in canine hepatic organoids. *Stem Cell Rep.* **2015**, *5*, 895–907. [CrossRef]
98. Shi, Y.; Inoue, H.; Wu, J.C.; Yamanaka, S. Induced pluripotent stem cell technology: a decade of progress. *Nat. Rev. Drug Discov.* **2016**, *16*, 115–130. [CrossRef]
99. Kimbrel, E.A.; Lanza, R. Current status of pluripotent stem cells: moving the first therapies to the clinic. *Nat. Rev. Drug Discov.* **2015**, *14*, 681–692. [CrossRef]
100. Mandai, M.; Watanabe, A.; Kurimoto, Y.; Hirami, Y.; Morinaga, C.; Daimon, T.; Fujihara, M.; Akimaru, H.; Sakai, N.; Shibata, Y.; et al. Autologous Induced Stem-Cell–Derived Retinal Cells for Macular Degeneration. *N. Engl. J. Med.* **2017**, *376*, 1038–1046. [CrossRef]
101. Lamm, N.; Ben-David, U.; Golan-Lev, T.; Storchová, Z.; Benvenisty, N.; Kerem, B. Genomic Instability in Human Pluripotent Stem Cells Arises from Replicative Stress and Chromosome Condensation Defects. *Cell Stem Cell* **2016**, *18*, 253–261. [CrossRef]
102. Iansante, V.; Mitry, R.R.; Filippi, C.; Fitzpatrick, E.; Dhawan, A. Human hepatocyte transplantation for liver disease: current status and future perspectives. *Pediatr. Res.* **2018**, *83*, 232–240. [CrossRef] [PubMed]
103. Matas, A.J.; Sutherland, D.E.R.; Steffes, M.W.; Michael Mauer, S.; Lowe, A.; Simmons, R.L.; Najarian, J.S. Hepatocellular transplantation for metabolic deficiencies: Decrease of plasma bilirubin in Gunn rats. *Science* **1976**, *192*, 892–894. [CrossRef] [PubMed]
104. Mito, M.; Kusano, M. Hepatocyte Transplantation in Man. *Cell Transplant.* **1993**, *2*, 65–74. [CrossRef]
105. Carpentier, A.; Tesfaye, A.; Chu, V.; Nimgaonkar, I.; Zhang, F.; Lee, S.B.; Thorgeirsson, S.S.; Feinstone, S.M.; Liang, T.J. Engrafted human stem cell-derived hepatocytes establish an infectious HCV murine model. *J. Clin. Investig.* **2014**, *124*, 4953–4964. [CrossRef]
106. Chen, Y.; Li, Y.; Wang, X.; Zhang, W.; Sauer, V.; Chang, C.J.; Han, B.; Tchaikovskaya, T.; Avsar, Y.; Tafaleng, E.; et al. Amelioration of Hyperbilirubinemia in Gunn Rats after Transplantation of Human Induced Pluripotent Stem Cell-Derived Hepatocytes. *Stem Cell Rep.* **2015**, *5*, 22–30. [CrossRef]
107. Tolosa, L.; Caron, J.; Hannoun, Z.; Antoni, M.; López, S.; Burks, D.; Castell, J.V.; Weber, A.; Gomez-Lechon, M.J.; Dubart-Kupperschmitt, A. Transplantation of hESC-derived hepatocytes protects mice from liver injury. *Stem Cell Res. Ther.* **2015**, *6*, 1–17. [CrossRef]
108. Nagamoto, Y.; Takayama, K.; Ohashi, K.; Okamoto, R.; Sakurai, F.; Tachibana, M.; Kawabata, K.; Mizuguchi, H. Transplantation of a human iPSC-derived hepatocyte sheet increases survival in mice with acute liver failure. *J. Hepatol.* **2016**, *64*, 1068–1075. [CrossRef]
109. Rashidi, H.; Luu, N.T.; Alwahsh, S.M.; Ginai, M.; Alhaque, S.; Dong, H.; Tomaz, R.A.; Vernay, B.; Vigneswara, V.; Hallett, J.M.; et al. 3D human liver tissue from pluripotent stem cells displays stable phenotype in vitro and supports compromised liver function in vivo. *Arch. Toxicol.* **2018**, *92*, 3117–3129. [CrossRef]

110. Lorvellec, M.; Scottoni, F.; Crowley, C.; Fiadeiro, R.; Maghsoudlou, P.; Pellegata, A.F.; Mazzacuva, F.; Gjinovci, A.; Lyne, A.M.; Zulini, J.; et al. Mouse decellularised liver scaffold improves human embryonic and induced pluripotent stem cells differentiation into hepatocyte-like cells. *PLoS ONE* **2017**, *12*, 1–23. [CrossRef]
111. Minami, T.; Ishii, T.; Yasuchika, K.; Fukumitsu, K.; Ogiso, S.; Miyauchi, Y.; Kojima, H.; Kawai, T.; Yamaoka, R.; Oshima, Y.; et al. Novel hybrid three-dimensional artificial liver using human induced pluripotent stem cells and a rat decellularized liver scaffold. *Regen. Ther.* **2019**, *10*, 127–133. [CrossRef]
112. Blackford, S.J.I.; Ng, S.S.; Segal, J.M.; King, A.J.F.; Austin, A.L.; Kent, D.; Moore, J.; Sheldon, M.; Ilic, D.; Dhawan, A.; et al. Validation of Current Good Manufacturing Practice Compliant Human Pluripotent Stem Cell-Derived Hepatocytes for Cell-Based Therapy. *Stem Cells Transl. Med.* **2019**, *8*, 124–137. [CrossRef] [PubMed]
113. Takayama, K.; Akita, N.; Mimura, N.; Akahira, R.; Taniguchi, Y.; Ikeda, M.; Sakurai, F.; Ohara, O.; Morio, T.; Sekiguchi, K.; et al. Generation of safe and therapeutically effective human induced pluripotent stem cell-derived hepatocyte-like cells for regenerative medicine. *Hepatol. Commun.* **2017**, *1*, 1058–1069. [CrossRef] [PubMed]
114. Nie, Y.Z.; Zheng, Y.W.; Ogawa, M.; Miyagi, E.; Taniguchi, H. Human liver organoids generated with single donor-derived multiple cells rescue mice from acute liver failure. *Stem Cell Res. Ther.* **2018**, *9*, 1–12. [CrossRef] [PubMed]
115. Rowe, R.G.; Daley, G.Q. Induced pluripotent stem cells in disease modelling and drug discovery. *Nat. Rev. Genet.* **2019**, *1*. [CrossRef]
116. Avior, Y.; Sagi, I.; Benvenisty, N. Pluripotent stem cells in disease modelling and drug discovery. *Nat. Rev. Mol. Cell Biol.* **2016**, *17*, 170–182. [CrossRef]
117. Sterneckert, J.L.; Reinhardt, P.; Schöler, H.R. Investigating human disease using stem cell models. *Nat. Rev. Genet.* **2014**, *15*, 625–639. [CrossRef]
118. Nie, Y.Z.; Zheng, Y.W.; Miyakawa, K.; Murata, S.; Zhang, R.R.; Sekine, K.; Ueno, Y.; Takebe, T.; Wakita, T.; Ryo, A.; et al. Recapitulation of hepatitis B virus–host interactions in liver organoids from human induced pluripotent stem cells. *EBioMedicine* **2018**, *35*, 114–123. [CrossRef]
119. Kisseleva, T.; Brenner, D.A. The Crosstalk between Hepatocytes, Hepatic Macrophages, and Hepatic Stellate Cells Facilitates Alcoholic Liver Disease. *Cell Metab.* **2019**, *30*, 850–852. [CrossRef]
120. Choi, W.M.; Kim, H.H.; Kim, M.H.; Cinar, R.; Yi, H.S.; Eun, H.S.; Kim, S.H.; Choi, Y.J.; Lee, Y.S.; Kim, S.Y.; et al. Glutamate Signaling in Hepatic Stellate Cells Drives Alcoholic Steatosis. *Cell Metab.* **2019**, *30*, 877–889.e7. [CrossRef]
121. Miranda, C.C.; Fernandes, T.G.; Pinto, S.N.; Prieto, M.; Diogo, M.M.; Cabral, J.M.S. A scale out approach towards neural induction of human induced pluripotent stem cells for neurodevelopmental toxicity studies. *Toxicol. Lett.* **2018**, *294*, 51–60. [CrossRef]
122. Sayed, N.; Liu, C.; Wu, J.C. Translation of Human-Induced Pluripotent Stem Cells from Clinical Trial in a Dish to Precision Medicine. *J. Am. Coll. Cardiol.* **2016**, *67*, 2161–2176. [CrossRef] [PubMed]
123. Knowlton, S.; Tasoglu, S. A Bioprinted Liver-on-a-Chip for Drug Screening Applications. *Trends Biotechnol.* **2016**, *34*, 681–682. [CrossRef]
124. Siramshetty, V.B.; Nickel, J.; Omieczynski, C.; Gohlke, B.O.; Drwal, M.N.; Preissner, R. WITHDRAWN—A resource for withdrawn and discontinued drugs. *Nucleic Acids Res.* **2016**, *44*, D1080–D1086. [CrossRef] [PubMed]
125. Carpentier, A.; Nimgaonkar, I.; Chu, V.; Xia, Y.; Hu, Z.; Liang, T.J. Hepatic differentiation of human pluripotent stem cells in miniaturized format suitable for high-throughput screen. *Stem Cell Res.* **2016**, *16*, 640–650. [CrossRef] [PubMed]
126. Choi, S.M.; Kim, Y.; Shim, J.S.; Park, J.T.; Wang, R.H.; Leach, S.D.; Liu, J.O.; Deng, C.; Ye, Z.; Jang, Y.Y. Efficient drug screening and gene correction for treating liver disease using patient-specific stem cells. *Hepatology* **2013**, *57*, 2458–2468. [CrossRef]
127. Cayo, M.A.; Mallanna, S.K.; Di Furio, F.; Jing, R.; Tolliver, L.B.; Bures, M.; Urick, A.; Noto, F.K.; Pashos, E.E.; Greseth, M.D.; et al. A Drug Screen using Human iPSC-Derived Hepatocyte-like Cells Reveals Cardiac Glycosides as a Potential Treatment for Hypercholesterolemia. *Cell Stem Cell* **2017**, *20*, 478–489.e5. [CrossRef]
128. Jing, R.; Corbett, J.L.; Cai, J.; Beeson, G.C.; Beeson, C.C.; Chan, S.S.; Dimmock, D.P.; Lazcares, L.; Geurts, A.M.; Lemasters, J.J.; et al. A Screen Using iPSC-Derived Hepatocytes Reveals NAD+ as a Potential Treatment for mtDNA Depletion Syndrome. *Cell Rep.* **2018**, *25*, 1469–1484.e5. [CrossRef]

129. Medine, C.N.; Lucendo-Villarin, B.; Storck, C.; Wang, F.; Szkolnicka, D.; Khan, F.; Pernagallo, S.; Black, J.R.; Marriage, H.M.; Ross, J.A.; et al. Developing high-fidelity hepatotoxicity models from pluripotent stem cells. *Stem Cells Transl. Med.* **2013**, *2*, 505–509. [CrossRef]
130. Sirenko, O.; Hancock, M.K.; Hesley, J.; Hong, D.; Cohen, A.; Gentry, J.; Carlson, C.B.; Mann, D.A. Phenotypic characterization of toxic compound effects on liver spheroids derived from ipsc using confocal imaging and three-dimensional image analysis. *Assay Drug Dev. Technol.* **2016**, *14*, 381–394. [CrossRef]
131. Ware, B.R.; Berger, D.R.; Khetani, S.R. Prediction of drug-induced liver injury in micropatterned co-cultures containing iPSC-derived human hepatocytes. *Toxicol. Sci.* **2015**, *145*, 252–262. [CrossRef]
132. Ronaldson-Bouchard, K.; Vunjak-Novakovic, G. Organs-on-a-Chip: A Fast Track for Engineered Human Tissues in Drug Development. *Cell Stem Cell* **2018**, *22*, 310–324. [CrossRef] [PubMed]
133. Zhang, B.; Korolj, A.; Lai, B.F.L.; Radisic, M. Advances in organ-on-a-chip engineering. *Nat. Rev. Mater.* **2018**, *3*, 257–278. [CrossRef]
134. Bhatia, S.N.; Ingber, D.E. Microfluidic organs-on-chips. *Nat. Biotechnol.* **2014**, *32*, 760–772. [CrossRef]
135. Wang, Y.; Wang, H.; Deng, P.; Chen, W.; Guo, Y.; Tao, T.; Qin, J. In situ differentiation and generation of functional liver organoids from human iPSCs in a 3D perfusable chip system. *Lab Chip* **2018**, *18*, 3606–3616. [CrossRef]
136. Miranda, C.C.; Fernandes, T.G.; Diogo, M.M.; Cabral, J.M.S. Towards multi-organoid systems for drug screening applications. *Bioengineering* **2018**, *5*, 49. [CrossRef] [PubMed]
137. Chen, W.L.K.; Edington, C.; Suter, E.; Yu, J.; Velazquez, J.J.; Velazquez, J.G.; Shockley, M.; Large, E.M.; Venkataramanan, R.; Hughes, D.J.; et al. Integrated gut/liver microphysiological systems elucidates inflammatory inter-tissue crosstalk. *Biotechnol. Bioeng.* **2017**, *114*, 2648–2659. [CrossRef]
138. Edington, C.D.; Chen, W.L.K.; Geishecker, E.; Kassis, T.; Soenksen, L.R.; Bhushan, B.M.; Freake, D.; Kirschner, J.; Maass, C.; Tsamandouras, N.; et al. Interconnected Microphysiological Systems for Quantitative Biology and Pharmacology Studies. *Sci. Rep.* **2018**, *8*, 1–18. [CrossRef]

MDPI
St. Alban-Anlage 66
4052 Basel
Switzerland
Tel. +41 61 683 77 34
Fax +41 61 302 89 18
www.mdpi.com

Bioengineering Editorial Office
E-mail: bioengineering@mdpi.com
www.mdpi.com/journal/bioengineering

www.ingramcontent.com/pod-product-compliance
Lightning Source LLC
LaVergne TN
LVHW070854170726
843515LV00005B/645

* 9 7 8 3 0 3 9 4 3 0 3 8 3 *